Lecture Notes in Mathematics

A collection of informal reports and seminars
Edited by A. Dold, Heidelberg and B. Eckmann, Zürich

308

Donald Knutson

Columbia University in the City of New York,
New York, NY/USA

λ-Rings and the Representation Theory of the Symmetric Group

T0218474

Springer-Verlag
Berlin · Heidelberg · New York 1973

AMS Subject Classifications (1970): 13A99, 20-02, 20C30

ISBN 3-540-06184-3 Springer-Verlag Berlin · Heidelberg · New York
ISBN 0-387-06184-3 Springer-Verlag New York · Heidelberg · Berlin

Offsetdruck: Julius Beltz, Hemsbach/Bergstr.

CONTENTS

Introduction 1

Chapter I: λ-Rings 5

1. The Definition of λ-Ring 5
2. General Constructions and Definitions on λ-Rings 15
3. Symmetric Functions 28
4. Adams Operations 46

Chapter II: The Representation Theory of Groups . . 59

1. The Representation Ring of a Finite Group . . 60
2. Irreducible Representations and Schur's Lemma . 76
3. Characters 81
4. Permutation Representations and the Burnside Ring 104
5. The Group Algebra Approach 115

Chapter III: The Fundamental Theorem 124

1. The Fundamental Theorem of the Representation Theory of the Symmetric Group 124
2. Complements and Corollaries 137
3. Schur Functions and the Frobenius Character Formula 155
4. Methods of Calculation. Young Diagrams . . 166

Bibliography 194

Index of Notation 198

Index 200

INTRODUCTION

These are notes from a seminar given at Columbia University in the year 1971-72. The object is to prove what is herein christened the Fundamental Theorem of the Representation Theory of the Symmetric Group. This theorem states that there is an isomorphism between a ring constructed by composing in a certain way all the representation rings of the symmetric groups S_n, and the ring of all symmetric polynomials in an infinite number of variables.

This isomorphism has been more or less known since the origins of the subject with Frobenius around 1900. But for various reasons it has been expressed (e.g., in Weyl's *Classical Groups*) through its relation to the representation theory of the general linear group. The isomorphism in its pure form seems not to have appeared until Atiyah's *Power Operations in K-Theory*, where a "dual" version is given in a brief introductory section.

The main technical tool is the notion of λ-ring, first introduced by Grothendieck in 1956 ([18]) in an algebraic-geometric context, and later used in group theory by Atiyah and Tall ([4]).

The notion of λ-ring is built upon the classical Fundamental Theorem of Symmetric Functions:

<u>Theorem</u>: Let $n \geq 1$ and $f(X_1, X_2, \ldots, X_n)$ be a polynomial with integer coefficients having the property that, for each permutation $\sigma \in S_n$, $f(X_{\sigma(1)}, X_{\sigma(2)}, \ldots, X_{\sigma(n)}) \equiv f(X_1, X_2, \ldots, X_n)$. ("f is symmetric".) Then there is a unique polynomial $F(a_1, a_2, \ldots, a_n)$, in variables $a_1, a_2, \ldots, a_n$, and with integer coefficients, so that, substituting

$$
\begin{aligned}
a_1 &= X_1 + X_2 + \ldots + X_n \\
a_2 &= X_1X_2 + X_1X_3 + X_2X_3 + \ldots + X_{n-1}X_n \\
&\vdots \\
a_i &= \sum_{j_1 < \ldots < j_i} X_{j_1} X_{j_2} \ldots X_{j_i} \\
&\vdots \\
a_n &= X_1 X_2 \ldots X_n
\end{aligned}
$$

(the "elementary symmetric functions" of $X_1, \ldots, X_n$) there is a polynomial identity

$$f(X_1, X_2, \ldots, X_n) \equiv F(a_1, a_2, \ldots, a_n) \quad .$$

(For a <u>proof</u>, and some applications, see van der Waerden [46]).

Much of classical algebra is built on this theorem. One is given a symmetric function of some variables - e.g., the resultant of two polynomials - and asked to compute the associated F.

A λ-ring R is a commutative ring with identity, and with operations $\lambda^n: R \longrightarrow R$, $n=0,1,\ldots$. Given $r \in R$, and any symmetric function $f(X_1,\ldots,X_n)$ of variables $X_1,\ldots,X_n$, an element $f(X_1,\ldots,X_n)(r) \in R$ is defined by, first find the polynomial $F(a_1,\ldots,a_n)$ associated to f, and then evaluate $F(\lambda^1 r, \lambda^2 r, \ldots, \lambda^n r)$. Axioms are given describing the composites $\lambda^n(r_1+r_2)$, $\lambda^n(r_1 r_2)$, and $\lambda^n(\lambda^m(r))$, and specifying $\lambda^0(r)=1$, $\lambda^1(r)=r$.

In Chapter I, we study λ-rings. Chapter II is a discussion, from scratch, of the representation theory of finite groups. The two are combined in Chapter III for the Fundamental Theorem.

Several relevant topics have been omitted here and will, we hope, appear in a more permanent version of this work in the future. One notable omission is the algorithm for multiplying Schur functions and its application to the classical Schubert Calculus.

One object of these notes is to present all of this theory from an elementary point of view. Beyond the standard Birkhoff/MacLane/van der Waerden ([7], [46]) algebra background, nothing is assumed (except occasionally a little bit of category theory). This is our main reason for including a complete discussion of the representation theory of groups.

It will be clear to many people how much we owe to various sources. Especially, this applies to Chapter II vis-a-vis Serre's

<ins>Representations</ins> <ins>Lineares</ins> <ins>des</ins> <ins>Groupes</ins> <ins>Finis</ins>, his exposition being slick enough that it is difficult to make any improvement.

While working on these notes, I was supported by the National Science Foundation (Contracts GP-21341, GP-33019x), and by Columbia University. I was supported mathematically by conversations with a large number of people at Columbia and elsewhere. Here, especially, I want to thank several people in the original seminar - Ron Proulx, Evelyn Boorman, and Mel Brender - for putting up with, and helping correct, my first version of these ideas.

D. Knutson

Columbia University

New York City

August 1972

CHAPTER I : λ-RINGS

1. The Definition of λ-ring

A λ-ring R is a commutative ring with identity on which certain additional unary operations are defined, $\lambda^n : R \longrightarrow R$, $n=0,1,\ldots$, satisfying certain axioms. The simplest example is the ring of integers $\mathbb{Z}$, for which $\lambda^n(m) = \binom{m}{n}$, the binomial coefficient. The general definition is somewhat complicated, however, and will be best understood by first analyzing one manifestation of the ring $\mathbb{Z}$, its appearance in the simplest example of K-theory. This is the construction of $\mathbb{Z}$ as the ring of finite dimensional vector spaces over a given field F.

Let F be a given field (the real numbers will do). Let $\mathcal{C}$ be the class of all finite dimensional vector spaces over F. Isomorphism of vector spaces gives an equivalence relation on $\mathcal{C}$, and we write $\overline{\mathcal{C}}$ for the set of equivalence classes, writing $[V]$ for the isomorphism class of V. Since two vector spaces are isomorphic if and only if they have the same dimension and the range of possible dimensions is exactly the set of non-negative integers, $\overline{\mathcal{C}}$ can be identified with this set, by $[V] \rightsquigarrow \dim V$.

In $\overline{\mathcal{C}}$ there is an operation of addition: we define $[V_1] = [V_2] + [V_3]$ if $V_1 = V_2 \oplus V_3$ (the direct sum). Additive inverses are introduced by enlarging $\overline{\mathcal{C}}$ to the set K(F): we define K(F) to be the set consisting of all expressions $[V_1] - [V_2]$, $[V_1], [V_2] \in \overline{\mathcal{C}}$, modulo the equivalence relation:

$$[V_1] - [V_2] = [V_1'] - [V_2'] \text{ if and only if } [V_1] + [V_2'] = [V_1'] + [V_2].$$

Taking $[V] = [V] - [0]$, $\mathcal{C}$ is a subset of K(F).

It is easy to show that K(F) is an abelian group. Indeed, identifying $\mathcal{C}$ with $\{0,1,2,..\}$, K(F) is isomorphic to $\mathbb{Z}$, as an additive group, and this is just the usual construction of $\mathbb{Z}$ from the positive integers. An element of K(F) corresponding to a positive integer can be written as $[V] - [0]$, for some vector space V, and thus thought of as an actual vector space. An element of K(F) corresponding to a negative integer is a "virtual vector space".

The tensor product $V_1 \otimes V_2$ of vector spaces introduces a multiplication in K(F), which under the isomorphism $K(F) \cong \mathbb{Z}$, corresponds to the usual multiplication of integers. Thus K(F) is a commutative ring with identity.

The exterior powers $\Lambda^d V$ of a vector space V now give rise to operations on K(F) corresponding to the formation of binomial coefficients $\binom{\dim V}{d}$ in $\mathbb{Z}$. Recall that if $\{v_1, v_2, \ldots, v_n\}$ is a basis for V, then one basis for $\Lambda^d V$ is the set of d-fold wedge products $\{ v_{i_1} \wedge v_{i_2} \wedge \cdots \wedge v_{i_d},\ 1 \le i_1 < \ldots < i_d \le n\}$, and there are $\binom{n}{d}$ such expressions if $0 < d \le n$, none such if $d > n$. By fiat we take $\Lambda^0 V = F$.

<u>Proposition:</u> i) $\Lambda^1 V = V$

ii) If $W = U \oplus V$, then for all positive integers n,

$$\Lambda^n W = \bigoplus_{i=0}^{n} (\Lambda^i U \otimes \Lambda^{n-i} V) \quad .$$

<u>Proof:</u> i) is obvious.

In ii), let $\{u_1,\ldots,u_{d_1}\}$ be a basis of U, and $\{v_1,\ldots,v_{d_2}\}$ be a basis of V, so that one basis of $W = U \oplus V$ is $\{u_1,\ldots,u_{d_1},v_1,\ldots,v_{d_2}\}$. A basis of $\wedge^n W$ is made up of symbols which are n-fold wedges of basis elements of W, and in any such wedge, changing the order of the elements just multiplies the wedge by ± 1. Hence one basis of $\wedge^n W$ consists of elements of the form $u_{i_1}\wedge\ldots\wedge u_{i_q}\wedge v_{j_1}\wedge\ldots\wedge v_{j_{n-q}}$, $1 \leq i_1 < \ldots < i_q \leq d_1$, $1 \leq j_i < \ldots < j_{n-q} \leq d_2$, $1 \leq q \leq n$. But such an element forms part of a basis of $\wedge^q U \oplus \wedge^{n-q} V$. ∎

Looking at the dimensions involved in this proposition, we have the identity

$$\binom{d_1+d_2}{n} = \sum_{i+j=n} \binom{d_1}{i}\binom{d_2}{j}$$

Thus the $\wedge^n$ operations induce on K(F) the structure of pre-λ-ring in the sense of the following definition.

<u>Definition</u>: A pre-λ-ring is a commutative ring R with identity 1, with a series of operations $\lambda^i : R \longrightarrow R$, $i=0,1,\ldots$, satisfying for all $x,y \in R$,

1) $\lambda^0(x) = 1$
2) $\lambda^1(x) = x$
3) $\lambda^n(x+y) = \sum_{i=0}^{n} \lambda^i(x)\lambda^{n-i}(y)$.

There is an equivalent definition. For $x \in R$, consider the formal power series in the variable t:

$$\lambda_t(x) = \lambda^0(x) + \lambda^1(x)t + \lambda^2(x)t^2 + \ldots$$

Then the requirements are that $\lambda^0(x)=1$, $\lambda^1(x)=x$, and $\lambda_t(x+y) = \lambda_t(x)\lambda_t(y)$. In this form it is evident that $\lambda_t(0)=1$ and $\lambda_t(-x)= 1/\lambda_t(x)$.

As we have seen, $\mathbb{Z}$ is a λ-ring, taking $\lambda_t(n) = (1+t)^n$. This is not the only pre-λ-ring structure on $\mathbb{Z}$. Given any power series with integer coefficients of the form $1 + t + n_2t^2 + n_3t^3 + . \quad .$ we can define, for all n, $\lambda_t(n) = (1 + t + n_2t^2 + . \ .)^n$. But with the definition later of λ-ring, we will rule out all λ-structures on $\mathbb{Z}$ except the original one above.

On the real numbers $\mathbb{R}$, we can define a pre-λ-structure by taking, for $r \in R$, $\lambda_t(r) = (1+t)^r$, expanding by the binomial theorem. Another pre-λ-structure is obtained by taking $\lambda_t(r)= e^{tr}$. (Again, the first structure will later pass muster as a λ-ring, but the second won't.)

Let R be a pre-λ-ring. An element $x \in R$ is <u>of finite degree n</u> if $\lambda_t(x)$ is a polynomial of degree n, or in other words, if $\lambda^m(x)=0$, for all $m > n$, but $\lambda^n(x) \neq 0$. R is <u>finitary</u> if each element of R is a difference of elements of finite degree. (For example, $\mathbb{Z}$ under $\lambda_t(n)=(1+t)^n$ is finitary, bur $\mathbb{R}$ under the same rule isn't.)

In either $\mathbb{Z}$ or $\mathbb{R}$, taking $\lambda_t(x)=(1+t)^x$, we have

$$\lambda^n(x) = \frac{x(x-1)(x-2)..(x-n+1)}{n(n-1)(n-2)..(1)} = \left(\frac{x}{n}\right)\cdot\left(\frac{(x-1)(x-2)..(x-n+1)}{(n-1)(n-2)..(1)}\right) = \left(\frac{x}{n}\right)\lambda^{n-1}(x-1).$$

In general, we say an element in a pre-λ-ring R is of <u>binomial type</u> if $n\lambda^n(x)=x\lambda^{n-1}(x-1)$ for all $n>0$. R is <u>binomial</u> if every element in R is of binomial type. Thus both $\mathbb{R}$ and $\mathbb{Z}$ are binomial under $\lambda_t(x)=(1+t)^x$, but not under any of the other pre-λ-ring structures mentioned above.

It turns out surprisingly that most of the interesting λ-rings are not binomial! In particular, for each finite group G, the representation ring of G (which we shall study in detail in the next chapter) is a λ-ring and is binomial if and only if G is the trivial group, For a specific example, the permutation group S_3 gives a ring $R(S_3)$ which as an abelian group is free on three generators. One can choose these three generators to be 1, a, and b, where 1 is the identity element, $a^2=1$, $b^2=1+a+b$, $ab=b$, $\lambda_t(1)=1+t$, $\lambda_t(a)=1+at$, and $\lambda_t(b)=1+bt+at^2$. In particular, $2\lambda^2(b)=2a$, while $b\lambda^{2-1}(b-1) = b\lambda^1(b-1)=b(b-1)=b^2-b= (1+a+b)-b = 1+a \neq 2a$, so $R(S_3)$ isn't binomial.

The definition of pre-λ-ring involves an expression for $\lambda^n(x+y)$. The definition of λ-ring involves also expressions for $\lambda^n(xy)$ and $\lambda^n(\lambda^m(x))$. We now derive what these expressions ought to be. Suppose R is a pre-λ-ring and $x\in R$ is a sum of elements of degree 1: $x= x_1 + x_2 + . . + x_n$, $\lambda_t(x_i) = 1+x_it$. Then

$$1 + \lambda^1(x)t + \lambda^2(x)t^2 + \ldots = \lambda_t(x) = \sum_{i=1}^{n} \lambda_t(1+x_i t)$$

$$= 1 + (x_1+x_2+..+x_n)t + (x_1x_2+x_1x_3+x_2x_3+..+x_{n-1}x_n)t^2 + \ldots + (x_1x_2\ldots x_n)t^n .$$

Thus $\lambda^q(x)$ is just the qth elementary symmetric function of the x_i's.

Now suppose $x = x_1+\ldots x_n$, and $y=y_1+\ldots,y_m$ are two elements of R which both happen to be expressible as the sum of elements x_i, y_j of degree 1. Furthermore, suppose that in R, the product of elements of degree 1 is again of degree 1. Then $xy = (\sum x_i)(\sum y_j) = \sum_{i,j} x_i y_j$, by assumption a sum of elements of degree 1. Hence

$$\lambda_t(xy) = \prod_{i,j} (1 + x_i y_j t) .$$

Hence $\lambda^1(xy)=xy$. $\lambda^2(xy) = \sum (x_i y_j)(x_{i'} y_{j'})$, the sum being taken over all different pairs (i,j), (i',j'). Hence

$$\lambda^2(xy) = 2\sum_{\substack{i<i'\\ j<j'}} x_i x_{i'} y_j y_{j'} + \sum_{j<j'} x_i^2 y_j y_{j'} + \sum_{i<i'} x_i x_{i'} y_j^2$$

$$= 2\Big(\sum_{i<i'} x_i x_{i'}\Big)\Big(\sum_{j<j'} y_j y_{j'}\Big) + \Big(\sum x_i^2\Big)\Big(\sum_{j<j'} y_j y_{j'}\Big) + \Big(\sum_{i<i'} x_i x_{i'}\Big)\Big(\sum y_j^2\Big) .$$

But $\sum x_i^2 = (\sum x_i)^2 - 2\sum_{i<i'} x_i x_{i'} = x^2 - 2\lambda^2 x$. Ditto for y.

$$\text{Hence } \lambda^2(xy) = 2\lambda^2 x\,\lambda^2 y + (x^2 - 2\lambda^2 x)\,\lambda^2 y + (y^2 - 2\lambda^2 y)\,\lambda^2 x$$

$$= x^2\,\lambda^2 y + y^2\,\lambda^2 x - 2\lambda^2 x\,\lambda^2 y.$$

A tedious but similar calculation gives $\lambda^3(xy) = x^3\lambda^3y + y^3\lambda^3x +$ $+ 3\lambda^3x\lambda^3y + xy\lambda^2x\lambda^2y - 3x\lambda^2x\lambda^3y - 3y\lambda^2y\lambda^3x$.

Similarly, since $\lambda_t(x) = \prod_i (1+x_it)$,

$$\lambda_t(\lambda^q x) = \prod_{1 \leq i_1 < .. < i_q \leq n} (1 + (x_{i_1} x_{i_2} .. x_{i_q})t) \quad .$$

Hence $\lambda^2(\lambda^2(x)) = \lambda^3x\lambda^1x - \lambda^4x$, and $\lambda^2(\lambda^3(x)) = \lambda^6x - \lambda^5x\lambda^1x + \lambda^4x\lambda^2x$. Etc.

In the λ-ring $\mathbb{Z}$, each of the expressions calculated above gives rise to an identity involving binomial coefficients. Thus

$$\binom{xy}{2} = x^2\binom{y}{2} + y^2\binom{x}{2} - 2\binom{x}{2}\binom{y}{2}$$

and

$$\binom{\binom{x}{2}}{2} = \binom{x}{3}\binom{x}{1} - \binom{x}{4} \qquad \text{Etc.}$$

Thus the general pattern emerges. The λ-powers, $\lambda^n x$, of an element x, give the elementary symmetric functions of some variables x_i, and then (using the Fundamental Theorem on Symmetric Functions - see p.2 and [46]) any other symmetric function of the x_i's can be expressed in terms of the $\lambda^i x$ in a unique fashion. In particular, $\lambda^n(xy)$ and $\lambda^n(\lambda^m(x))$ are both symmetric functions of the variables involved, so have, uniquely by the Theorem, expressions in terms of the λ-powers of x and y.

Of course, all this holds a priori only if x and y are each expressible as a finite sum of elements of degree 1, and if the product of two elements of degree one is again of degree one. But we can use this special case to figure out what the identities involved are, and then build all the identities into the formal definition of λ-ring. Hence given a pre-λ-ring which is known to be a λ-ring, any calculation of $\lambda^5(xy)$ or $\lambda^7(\lambda^3(x))$, say, can be carried out by just pretending that x is the sum of a finite number of elements of degree 1: $x = x_1+x_2+..+x_n$ (with $n \geq 5$ in the first example and $n \geq 7\cdot 3 = 21$ in the second) calculating the result as a symmetric function in the x_i's, and then passing to a polynomial in the λ-powers of x. We now give the formal definition of λ-ring in such a way that the validitiy of this procedure is built in.

Let $\xi_1, \xi_2, \ldots, \xi_q; \eta_1, \eta_2, \ldots, \eta_r$ be indeterminates. Define s_i and σ_j by

$$(1 + s_1 t + s_2 t^2 + ..) = \prod_i (1 + \xi_i t)$$

$$(1 + \sigma_1 t + \sigma_2 t^2 + ..) = \prod_j (1 + \eta_j t)$$ - in other words, the

s_i's and σ_j's are the elementary symmetric functions of the ξ_i's and η_j's. Let $P_n(s_1, s_2, \ldots, s_n; \sigma_1, \sigma_2, \ldots, \sigma_n)$ be the coefficient of

t^n in $\prod_{i,j} (1 + \xi_i \eta_j t)$ and $P_{nd}(s_1, s_2, \ldots, s_{nd})$ be the coefficient of t^n in the product $\prod_{1 \leq i_1 < \ldots < i_d \leq q} (1 + \xi_{i_1} \xi_{i_2} \ldots \xi_{i_d} t)$. By the Fundamental Theorem on Symmetric Functions, P_n and P_{nd} are polynomials with integer coefficients, and are independent of q and r as long as $q \geq n$ and $r \geq n$ in the first case, and $q \geq nd$ in the second.

since P_n and P_{nd} have integer coefficients, they are well-defined over any ring with identity, and so are sometimes referred to as <u>universal</u> polynomials.

<u>Main Definition</u>: A λ-ring R is a pre-λ-ring in which

1) $\lambda_t(1) = 1+t$

2) For all x,y in R, $n \geq 0$, $\lambda^n(xy) = P_n(\lambda^1 x, \lambda^2 x, \ldots, \lambda^n x, \lambda^1 y, \ldots \lambda^n y)$

3) For all x in R, and $n, m \geq 0$,

$$\lambda^m(\lambda^n(x)) = P_{mn}(\lambda^1 x, \ldots, \lambda^{mn} x)$$

Note the requirement 1) already rules out the exotic λ-structures given on $\mathbf{Z}$ and $\mathbf{R}$ above. 1) also requires that R is of characteristic zero, since for any integer m, $\lambda_t(m) = \lambda_t(\underbrace{1 + 1 + \ldots + 1}_{m \text{ times}})$

$= \lambda_t(1)^m = (1+t)^m = 1 + \ldots + t^m$. This last polynomial is not zero in any ring. Hence in any λ-ring $\underbrace{1 + 1 + 1 + \ldots + 1}_{m \text{ times}} \neq 0$.

In any λ-ring, the procedure for reducing an expression like $\lambda^3(x\lambda^2 y)$, say, in terms of the λ-powers of x and y can be carried through as indicated above. Indeed, that procedure amounts to little more than in each case calculating the universal polynomial P_n or P_{nm} from scratch and plugging in values for the s_i, σ_j

Unfortunately, from this definition, it is not so clear

how to prove that anything in particular, ($\mathbb{Z}$, say) is a λ-ring! In the next section, we will give a canonical construction for producing lots of λ-rings, and use it to show that $\mathbb{Z}$, under $\lambda_t(n) = (1+t)^n$ is one. The most convenient criterion, though, involves the Adams operations and is given on p. 49 .

2. General Constructions and Definitions on λ-rings

Let A be a ring, commutative with identity. Let $1 + A[[t]]^+$ denote the set of all formal power series $\sum_{n=0}^{\infty} a_n t^n$, with $a_n \in A$, for all n, and $a_0=1$. (As usual in the theory of formal power series, no questions of convergence arise - any infinite sequence of a_n is allowed, and the powers of t just serve to index the coefficients a_n.) Such power series, with first term 1, are called special.

We endow $1+A[[t]]^+$ with operations of addition "+", multiplication ".", and λ-powers "Λ^n" as follows: If $a = 1+a_1t+a_2t^2+..$, and $b = 1+b_1t+b_2t^2+..$, then a"+"b is the usual power series product of a and b:

$$a"+"b = 1 + (a_1+b_1)t + (a_2+a_1b_1+b_2)t^2 + (a_3+a_2b_1+a_1b_2+b_3)t^3 +..$$

The additive unit element "0" is then the power series $1+0t+0t^2+..$. The product a"."b is that power series $c=1+c_1t+c_2t^2+..$, with $c_n = P_n(a_1,a_2,..,a_n;b_1,b_2,..,b_n)$, where P_n is the polynomial defined above (page 12). In particular, given two power series whose coefficients are zero after the first term, say, $c = 1+c_1t$ and $d= 1+d_1t$, the product is simply $cd= 1+c_1d_1t$. The multiplicative unit "1" is the series 1+t. Finally, for an integer $i \geq 0$, $c="\Lambda^i"(a)$ is the power series $c=1+c_1t+c_2t^2+..$, where $c_n=P_{ni}(a_1,a_2,..,a_{ni})$, with P_{ni} the polynomial defined above (page 12).

Proposition: Under the operations defined above, $1+[[t]]^+$ is a pre-λ-ring.

Proof: Let us check some particular axiom of pre-λ-rings - say the distributive axiom: a"."(b"+"c) = (a"."b)"+"(a"."c). Each side is a power series in t, computed from a,b,c and to show the equality, it is sufficient to check that for each integer n, the coefficients of t^n on both sides agree. These coefficients are polynomials in the quantities $a_1,..,an,b_1,..,b_n,c_1,..,c_n$. Hence to check the identities it is sufficient to check them in the case where $a_1,...,c_n$ are algebraically independent. So let ζ_i,η_j,ζ_k be independent variables and let $1+a_1t+a_2t^2+..=\prod(1+\xi_it)$, $1+b_1t+b_2t^2+..=\prod(1+\eta_jt)$, and $1+c_1t+c_2t^2+..=\prod(1+\zeta_kt)$. The left side of the distributive axiom to be proved is then computed:

$$b"+"c = \prod(1+\eta_jt)\cdot\prod(1+\zeta_kt)$$

$$a"."(b"+"c) = \prod_{i,j}(1+\xi_i\eta_jt)\cdot\prod_{i,k}(1+\xi_i\zeta_kt)$$

(here the definition of P_n is used explicitly). This is exactly $\prod(1+\xi_i\eta_jt)$ "+" $\prod(1+\xi_i\zeta_kt)$ which is the right hand side. The other pre-λ-ring axioms follow similarly. ∎

Several facts are simple consequences of the definition of the operations in $1+A[[t]]^+$. Letting $a=\sum a_nt^n$, a is of degree n if and only if a is an nth degree polynomial: $a_m=0$, $m>n$, but $a_n\neq 0$. If a is of degree n, "Λ^n"a = $1+a_nt$. For any a, if $b=1+b_1t$ is of degree 1, a"."b = $\sum a_n(b_1t^n)$.

Let A be a pre-λ-ring. Consider the map

$$\lambda_t: A \longrightarrow 1+A[[t]]^+ \quad \text{defined by} \quad a \rightsquigarrow \lambda_t(a).$$

Proposition: A is a λ-ring if and only if λ_t is a homomorphism of pre-λ-rings.

Proof: By the definition of "+", λ_t is always additive, for any ring A. The operations "." and "Λ^n" have been exactly defined to make the proposition true.

Thus we have an alternate definition of λ-ring. With this it is somewhat easier to prove that certain rings are λ-rings.

Proposition: Let R be a pre-λ-ring in which

i) $\lambda_t(1) = 1+t$

ii) Each element $r \in R$ is expressible in the form $r = \sum \pm a_i$, a finite sum in which each summand is plus-or-minus an element a_i of degree 1

iii) The product of two elements of degree one is again of degree one.

Then R is a λ-ring.

Proof: We are given $\lambda_t: R \longrightarrow 1+R[[t]]^+$ additive, and $\lambda_t(1) =$ "1" $= 1+t$. We must check that $\lambda_t(xy) = \lambda_t(x)\lambda_t(y)$ and $\lambda_t(\lambda^n(x)) =$ "Λ^n"$(\lambda_t(x))$.

First we check the multiplicativity. Since x and y are each sums of $\pm$ elements of degree 1, and λ_t is additive, we can just do the case where x and y are of degree 1. Hence their product is of degree 1, which is to say, $\lambda_t(xy) = 1+xyt$. But this latter equals $(1+xt)"."(1+yt) = \lambda_t(x)"."\lambda_t(y)$.

The proof of $\lambda_t(\lambda^n(x)) = "\Lambda^n"(\lambda_t(x))$ is similar. Suppose the identity holds for x=a, b. We then compute $\lambda_t(\lambda^n(a+b)) =$

$$\lambda_t(\sum \lambda^i(a)\lambda^{n-i}(b)) = "\sum" \lambda_t(\lambda^i(a)\lambda^{n-i}(b)) =$$

$$"\sum"(\lambda_t(\lambda^i(a))"."\lambda_t(\lambda^{n-i}(b))) = "\sum"("\Lambda^i"(\lambda_t(a)"."\text{"}\Lambda^{n-i}\text{"}(\lambda_t(b)))$$

$$= "\Lambda^n"(\lambda_t(a)"+"\lambda_t(b)) = "\Lambda^n"(\lambda_t(a+b)).$$

So again, we can check the identity for x of degree one, which is straightforward. ∎

<u>Corollary</u>: $\mathbb{Z}$, under the λ-operations defined by $\lambda^n(m) = \binom{m}{n}$ is a λ-ring. ∎

(There is a converse to this proposition - the Splitting Principle ([4]) asserting every λ-ring can be imbedded inside a λ-ring of the type described in the proposition above.)

<u>Theorem:</u> (Grothendieck) For any commutative ring A with identity, the pre-λ-ring $1+A[[t]]^+$ is a λ-ring.

<u>Proof</u>: We already know that

$$"\Lambda^n"(1+t) = \begin{cases} "1" = 1+t & n=0,1 \\ "0" = 1 & n \geq 2 \end{cases}.$$

Thus we must show, for each $n \geq 1$, that "Λ^n"(x"."y) = P_n(x,"Λ^2"x,...,"Λ^n"x;y,"Λ^2"y,...,"Λ^n"y), and the same for iterated λ-powers. But as in the proof on page 16 , this again boils down to proving certain formal polynomial identities. To check each of these identities, we can assume that x and y are the "sum" of a finite number of elements "of degree 1" in $1+A[[t]]^+$. Hence as far as the things we have to check are concerned, we can just look at the elements of $1+A[[t]]^+$ which are sums and differences of elements of degree 1. The product of two elements of degree one is again of degree one in $1+A[[t]]^+$ so these elements form a sub-pre-λ-ring of $1+A[[t]]^+$. Since this sub-pre-λ-ring satisfies the hypotheses ot the previous proposition, it is a λ-ring, so the identities hold. ∎

A <u>map of λ-rings</u> $R_1 \longrightarrow R_2$, is a ring homomorphism $f:R_1 \longrightarrow R_2$, such that $\lambda^n(f(r))=f(\lambda^n(r))$ for all $r \in R_1$, and all $n \geq 0$. An <u>augmented</u> λ-ring is a λ-ring R together with a map of λ-rings $\varepsilon:R \longrightarrow \mathbb{Z}$.

The class of all λ-rings and maps of λ-rings form the <u>category of λ-rings</u>, which we denote by (λ-rings). The special power series ring construction provides a functor from the category of rings, (Rings), to (λ-rings): $A \rightsquigarrow 1+A[[t]]^+$. Let $U:(\lambda\text{-rings}) \longrightarrow (\text{Rings})$ be the "forgetful functor" assigning to each λ-ring R, its underlying ring R (with the λ-structure ignored).

<u>Proposition:</u> Let R be a ring and S be a λ-ring. Then there is a one-one correspondence between maps of λ-rings $\varphi:S \longrightarrow 1+R[[t]]^+$, and maps of rings $\psi:U(S) \longrightarrow R$.

<u>Proof</u>: We make use of two maps. The first is the map $f:1+R[[t]]^+ \longrightarrow R$, by $f(1+a_1t+a_2t^2+..)=a_1$, which is a map of rings. The second is the map of λ-rings, $\lambda_t:S \longrightarrow 1+S[[t]]^+$.

Now for each map of λ-rings $\varphi:S \longrightarrow 1+R[[t]]^+$, composing with f gives a map of rings $f\varphi:U(S) \longrightarrow R$. Conversely, a map of rings $\psi:S \longrightarrow R$ gives a map of λ-rings $\psi:1+S[[t]]^+ \longrightarrow 1+R[[t]]^+$, $\psi(1+s_1t+s_2t^2+..)=1+\psi(s_1)t+\psi(s_2)t^2+..$ Composing this with λ_t gives a map of λ-rings $\psi\lambda_t:S \longrightarrow 1+R[[t]]^+$. These two operations are inverse. ∎

From a categorical point of view*, this result is surprising.

*The reader who is not categorically inclined can skip this paragraph without loss of continuity.

It says that the forgetful functor (λ-rings) $\longrightarrow$ (Rings) has a <u>right</u> adjoint. By far the usual situation is that a forgetful functor, (e.g., the one from (Groups) to (sets) taking a group to its underlying set) has only a left adjoint (in that case, the construction of the free group on a set). Of course, our functor U also has a left adjoint (which the reader may construct). Hence it is both left and right exact. This fact can be used to carry out constructions in the category of λ-rings. For example, to define a product RxS of λ-rings, it must be the case by left exactness of U that if the usual categorical definition of RxS is representable in (λ-rings), then $U(R\times S)=U(R)\times U(S)$. Thus to construct a product of R and S, one only need find the appropriate λ-ring structure on the product of rings RxS. Sinilarly for tensor products. Motivated by these considerations, we make the following definitions.

Let R and S be λ-rings. We define the <u>product</u> or R and S, RxS, to be, as a set, the cartesian product of R and S. Ring operations are defined coordinate-wise: $(r_1,s_1)+(r_2,s_2)=(r_1+r_2,s_1+s_2)$ $(r_1,s_1)\cdot(r_2,s_2)=(r_1r_2,s_1s_2)$. We define $\lambda_t(r,s)=\lambda_t(r,0)\cdot\lambda_t(0,s)$,

and $\lambda_t(r,0) = (1,1) + \sum_{n \geq 1} (\lambda^n(r),0)t^n$, and similarly for $\lambda_t(0,s)$. The <u>tensor product</u> of two λ-rings is constructed by taking the tensor product $R\otimes S$ of the rings R and S, and defining $\lambda^n(a\otimes 1)=\lambda^n(a)\otimes 1$, $\lambda^n(1\otimes b)=1\otimes\lambda^n(b)$, and $\lambda^n(a\otimes b)=\lambda^n(a\otimes 1\cdot 1\otimes b)$. Similarly, the inverse limit of an inverse system of λ-rings, $\varprojlim R_i$, is obtained by taking the inverse limit as rings and imposing the obvious λ-structure. (The verification of all this is straightforward but rather tedious so we leave it to the reader.)

Proposition: Let R be a λ-ring and R[X] the ring of polynomials in one variable X over R. Then there is a unique structure of λ-ring on R[X] satisfying the three equivalent requirements:

i) X is of degree 1

ii) X^n is of degree 1 for all $n \geq 1$

iii) $\lambda^q(rX^n) = \lambda^q(r)X^{nr}$ for all integers $n,q \geq 0$, $r \in R$.

Furthermore, if R is augmented, R[X] may be augmented either by $\varepsilon(X) = 0$ or $\varepsilon(X) = 1$.

Proof: The equivalence of the three requirements on an element X in a λ-ring has been noted previously.(page 16) The uniqueness is clear: if $f(x) = \sum a_i x^i \in R[X]$ is any element, then for R[X] to be a λ-ring satisfying the requirements, it must be that

$\lambda_t(f(X)) = \lambda_t(\sum (a_i X^i)) = \prod \lambda_t(a_i X^i) = \prod \lambda_{X^i t}(a_i)$.

Taking this as the definition of $\lambda_t(f(X))$ in R[X], it is easy to see that $\lambda_t(f(X)+g(X)) = \lambda_t(f(X))\lambda_t(g(X)) = \lambda_t(f(X)) "+" \lambda_t(g(X))$.

To check the multiplicativity, it is sufficient to consider the case of the product of monomials. We only need show

i) $\lambda_t(rs) = \lambda_t(r) "." \lambda_t(s)$ $\quad r,s \in R$

ii) $\lambda_t(X^m X^n) = \lambda_t(X^m) "." \lambda_t(X^n)$ $\quad m,n \geq 0$

iii) $\lambda_t(rX^n) = \lambda_t(r) "." \lambda_t(X^n)$ $\quad r \in R,\ n \geq 0$.

i) is true by hypothesis and ii) and iii) follow easily from the definition given here of λ_t.

In checking that $\lambda_t(\lambda^n(f(X)))$="$\Lambda^n$"$(\lambda_t(f(X)))$, it is sufficient, since λ_t is additive and multiplicative, to show this for the two trivial cases: $f(X)=X$, and $f(X)=r$, r an element of R. Each of these is trivial.

In assigning an augmentation to X, the only condition is that $\varepsilon(X) \leq$ degree of $X = 1$. Say we wish to require $\varepsilon(X)=0$. Then the λ-ring homomorphism $R[X] \to R$, by $f(X) \rightsquigarrow f(0)$, composed with $\varepsilon: R \longrightarrow \mathbb{Z}$ gives an augmentation on $R[X]$ with $\varepsilon(X)=0$. For $\varepsilon(X)=1$, the first map is taken to be $f(X) \rightsquigarrow f(1)$. ∎

We now construct the free λ-ring on one generator. Let $\Omega_0 = \mathbb{Z}$, and Ω_n be constructed from Ω_{n-1} by adding an indeterminate ξ_n, as in the previous proposition. Thus $\Omega_n = \mathbb{Z}[\xi_1, \xi_2, \ldots, \xi_n]$, with $\lambda_t(\xi_i)=1+\xi_i t$, is a λ-ring, and for any $r > n$, there is a λ-ring homomorphism $\varphi_{rn}: \Omega_r \longrightarrow \Omega_n$, $\varphi(\xi_i) = \xi_i$ or 0, depending on whether $i \leq n$ or not.

Let $\Omega = \underleftarrow{\text{Lim}}\ \Omega_r$. (Recall this is the set of all sequences $(\ldots, a_n, a_{n-1}, \ldots, a_1, a_0)$ with $a_i \in \Omega_i$ and $\varphi_{n+i,n}(a_{n+1})=a_n$ for all n. Thus an element of Ω is an infinite power series in the infinite set of variables ξ_i, having the property that it reduces to a polynomial if all but a finite number of the variables ξ_i are set equal to zero.) As remarked above, (page 22) Ω is a λ-ring.

Let $a_n = a_n(\xi_1,\xi_2,\ldots,\xi_r) \in \Omega_r$ be the nth elementary symmetric function of the ξ's (equal by definition to zero if $n > r$) and $a_n \in \Omega$ be the associated element in the inverse limit. Then $\lambda^n(a_1)=a_n$, for all n (since this is true in each Ω_r) and the a_n's are algebraically independent (since any polynomial idetity involving some of $a_1,a_2,\ldots,a_r$, for example, would already give a polynomial identity involving $a_1(\xi_1,\ldots,\xi_r),\ldots,a_r(\xi_1,\ldots,\xi_r) \in \Omega_r$ and this is ruled out by the theorem of symmetric functions).

Let $\Lambda \subset \Omega$ be the sub-λ-ring generated by a_1. By the previous comments, Λ, as a ring, is a polynomial ring in an infinite number of indeterminates: $\Lambda = \mathbb{Z}[a_1,a_2,\ldots]$. Indeed any sub-$\lambda$-ring of Ω containing a_1 must contain each $a_n=\lambda^n(a_1)$, and so contain $\mathbb{Z}[a_1,a_2,\ldots]$. On the other hand, using the universal polynomials, we can evaluate any expression of the form $\lambda^n(f(a_1,a_2,\ldots))$, $f(a_1,a_2,\ldots) \in \mathbb{Z}[a_1,a_2,\ldots]$ as again a polynomial in the a_i with integer coefficients. Hence $\mathbb{Z}[a_1,a_2,\ldots]$ is a sub-λ-ring of Ω.

In the next section, and in Chapter Three, we will study intensively the free λ-ring on one generator. Here we wish just wish to mention a categorical application. A <u>natural operation</u> α

on the category of λ-rings is a natural transformation from the underlying set ~~identity~~ functor to itself. That is, we have an assignment to each λ-ring R, a map (of sets) $\alpha_R : R \longrightarrow R$ such that for any map of λ-rings $f:R \longrightarrow S$, $f\alpha_R = \alpha_S f : R \longrightarrow S$. Addition of natural operations is defined by $(\alpha_R + \beta_R)(r) = \alpha_R(r) + \beta_R(r)$ and similarly for multiplication and λ-operations.

<u>Proposition:</u> The set of natural operations is a λ-ring, and is isomorphic to the free λ-ring on one generator, Λ.

<u>Proof</u>: Let μ be an operation. Let $a_1 \in \Lambda$ be the generator of Λ and suppose $\mu_\Lambda(a_1) = f(a_1, a_2, \ldots) \in \Lambda$. For any λ-ring R, and $r \in R$, let $g: \Lambda \longrightarrow R$ be the unique λ-ring homomorphism with $g(a_1) = r$. Then $\mu_R(r) = \mu_R g(a_1) = g(\mu_\Lambda(a_1)) = g(f(a_1, a_2, \ldots)) = f(r, \lambda^2(r), \ldots)$. Conversely, given any element $f(a_1, a_2, \ldots) \in \Lambda$, the unique map $\Lambda \rightarrow \Lambda$, taking a_1 to $f(a_1, a_2, \ldots)$ extends uniquely to a natural operation. ∎

Hence given a natural operation, it is uniquely a polynomial in the λ-operations. To check that a given polynomial $f(\lambda^1, \lambda^2, \ldots)$ is equal to a given μ, one only need check $\mu(a_1) = f(a_1, a_2, \ldots)$. This being a proposed identity in $\Lambda \subset \Omega = \varprojlim \Omega_n$, it is sufficient to check in each Ω_n. This can be formally phrased as the

<u>Verification Principle</u>: If μ is a λ-ring operation, then μ is uniquely a polynomial in the λ-operations and for any particular polynomial $f(\lambda^1, \lambda^2, ..)$, to check that $\mu = f$, it is sufficient to check that $\mu = f$, operating on a sum $\xi_1 + \xi_2 + .. + \xi_r$ of elements of degree 1, for all $r > 0$.

One can similarly construct the free λ-ring on k generators, any $k \geq 1$ and discuss k-ary natural operations on the category of λ-rings, and formulate a similar verification principle for such.

This process can be generalized as follows. Let C be any category. By analogy with the case where C is the category of [redacted] finite groups, or the category of [redacted] compact Hausdorff spaces, we can define a <u>K-theory</u> to be a contravariant functor K_0 from C to the category of λ-rings. Given such, an <u>operation</u> in the K-theory K_0 is a natural transformation from the composite functor $C \longrightarrow (\lambda\text{-rings}) \longrightarrow (\text{Sets})$ to itself. The set of all such operations for a given C and K form a λ-ring, $Op(K_0)$, and there is a natural map of λ-rings $\Lambda \longrightarrow Op(K_0)$. Of course, in general this map need be neither one-one nor onto.

3. Symmetric Functions

Let Λ be the free λ-ring on one generator. Recall this means that Λ as a ring is a polynomial ring over $\mathbb{Z}$ in an infinite number ov variables: $\Lambda = \mathbb{Z}[a_1, a_2, \ldots]$, and $\lambda^n(a_1) = a_n$.

Λ was constructed as a subring of a λ-ring $\Omega = \varprojlim \Omega_n$, $\Omega_n = \mathbb{Z}[\xi_1, \xi_2, \ldots]$, and we can sum up the relation between the a's and the ξ's by the equation $\lambda_t(a_1) = \sum_{n=0}^{\infty} a_n t^n = \prod_{i=1}^{\infty} (1+\xi_i t)$. An element $f(a_1, a_2, \ldots, a_n) = F(\xi_1, \xi_2, \ldots)$ of Λ is called <u>isobaric of weight k</u> in the a_i's if as a function of the ξ_j's, it is homogeneous of degree k. Thus a monomial $a_1^{r_1} a_2^{r_2} \ldots a_n^{r_n}$ has weight $r_1+2r_2+3r_3+\ldots+nr_n$. Let Λ_n denote the set of all elements isobaric of weight n in Λ.

Λ_n is an abelian group under addition, since the sum of two elements of weight n is again of weight n. Since the set $\{a_1, a_2, \ldots\}$ forms a polynomial basis of Λ, the set of monomials $a_1^{r_1} a_2^{r_2} \ldots a_n^{r_n}$, (all n, all sequences $r_1, \ldots, r_n$ of non-negative integers) form an additive basis of Λ. Hence one basis of Λ_n consists of all monomials of the form $a_1^{r_1} a_2^{r_2} \ldots a_n^{r_n}$ with $\sum_i ir_i = n$. The number of such monomials is the number of partitions of the number n. Indeed, to each partition $\pi = 1^{r_1} 2^{r_2} \ldots n^{r_n}$ of n, we can associate the monomial $a_1^{r_1} a_2^{r_2} \ldots a_n^{r_n}$. This monomial will be denoted a_π. Thus Λ_n is a free abelian group with basis the set of all a_π, π a partition of n.

At this point it is convenient to introduce some general notation on partitions. Given a number n, a <u>partition</u> π of n is any sum $n = n_1+n_2+\ldots+n_k$, $n_i > 0$. If r_1 of the n's are equal to 1, r_2 are equal to 2, etc., this partition is denoted $\pi = 1^{r_1}2^{r_2}\ldots n^{r_n}$. Two partitions are equal if and only if the corresponding $r_1, r_2, \ldots$ are equal. Another common notation is to write the parts in decreasing order: $\pi = (n_1, n_2, \ldots, n_k)$, with $n_1 \geq n_2 \geq \ldots \geq n_k$. The diagram (<u>Ferrar</u>'s graph or <u>Young</u> diagram) associated with the partition $\pi = (n_1, n_2, \ldots, n_k)$ consists of n squares, arranged in k rows, with the ith row containing n_i of the squares, and all of the rows lined up at the left. Thus $\pi = (6,5,3,3,1,1,1)$ gives

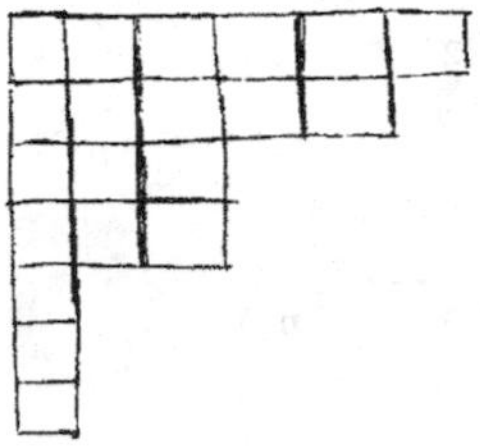

Given any partition π, we can draw its graph and let $l_1, l_2, \ldots, l_q$ be the lengths of the columns. The sequence $(l_1, l_2, \ldots, l_q)$ is also a partition of n, the <u>conjugate</u> partition of π, denoted π'. Its diagram is obtained by flipping the diagram for π along its diagonal. Thus for $\pi = (6,5,3,3,1,1,1)$, $\pi' = (7,4,4,2,2,1)$.

The set of all partitions of a number n is denoted $\Pi(n)$. (For a table of the size of $\Pi(n)$ for n=1,2,...,200, see [31]). For $\sigma \in \Pi(n)$, another common notation is $\sigma \vdash n$.

Now back to Λ_n. Since the weight of the product of two monomials of Λ is the sum of the weights, multiplication in Λ gives a map $\Lambda_n \times \Lambda_k \longrightarrow \Lambda_{n+k}$, for each n,k. Hence $\Lambda = \sum_{n \geq 0} \Lambda_n$ is a graded ring.

As shown above, each Λ_n is a free abelian group with a basis $\{a_\pi \mid \pi \vdash n\}$. But this is not the only"natural" basis. From the classical theory of symmetric functions, several other "natural" bases are also indicated.

The first of these is the set of <u>homogeneous power sums</u>. In terms of the variables ξ_i, we can define

$$h_1 = \sum \xi_i = a_1$$

$$h_2 = \sum_{i \leq j} \xi_i \xi_j = a_1^2 - a_2$$

$$\vdots$$

$$h_n = \sum_{i_1 \leq i_2 \leq \ldots \leq i_n} \xi_{i_1} \xi_{i_2} \cdots \xi_{i_n}$$

$$\vdots$$

This definition can also be written as follows: We have $\lambda_t(a_1) = \sum a_n t^n = \prod (1+\xi_i t)$. Thus

$$\frac{1}{\lambda_{-t}(a_1)} = \frac{1}{\prod_i (1-\xi_i t)} = \prod_i \left(\frac{1}{1-\xi_i t}\right) =$$

$$= \prod_i \left(\sum_n \xi_i^n t^n\right) = \sum_n h_n t^n .$$

Hence the a's and h's are related by the identity

$$\left(\sum_{n=0}^{\infty} a_n t^n\right)\left(\sum_{n=0}^{\infty} (-1)^n h_n t^n\right) = 1 \quad .$$

Equating coefficients of t^n gives

$$a_n - h_1 a_{n-1} + \cdots + (-1)^n h_n = 0 \qquad n \geq 1 \ .$$

Hence the set $\{ h_n \mid n \geq 0 \}$ also forms a polynomial basis of Λ, with each h_n of weight n. Given a partition $\pi = (1^{r_1} 2^{r_2} \ldots)$, we write $h_\pi = h_1^{r_1} h_2^{r_2} \ldots$. Then another basis for Λ_n as free abelian group is $\{ h_\pi \mid \pi \vdash n \}$.

In some cases, the natural way to describe the λ-structure on a λ-ring R is to specify not the operations λ^n, but rather the operations h_n. By the formulas above, this amounts to the same thing , since the λ^n's are expressed in terms of the h_n's. And, of course, the same comment would hold for any choice of bases of Λ_n, $n \geq 1$.

Another basis is given by the <u>monomial</u> <u>symmetric</u> <u>functions</u>. Let $\pi = (1^{r_1} 2^{r_2} \ldots)$ be a partition of a number n. We give two descriptions of the monomial symmetric function $\langle \pi \rangle$ associated to π. One is a formula

$$\langle \pi \rangle = \sum_{\substack{i_1 < i_2 < \ldots < i_{r_1} \\ j_1 < j_2 < \ldots < j_{r_2} \\ k_1 < k_2 < \ldots < k_{r_3} \\ \text{(all indices distinct)}}} \xi_{i_1} \xi_{i_2} \ldots \xi_{i_{r_1}} \xi_{j_1}^2 \xi_{j_2}^2 \ldots \xi_{j_{r_2}}^2 \xi_{k_1}^3 \ldots \xi_{k_{r_3}}^3 \ldots$$

For a more abstract definition, consider the set of all monomials $\xi_{i_1}^{r_1} \xi_{i_2}^{r_2} \ldots \xi_{i_k}^{r_k}$, all $r_i > 0$, $i_j > 0$ in Ω. The infinite symmetric group S_∞ acts on this set by, for $\sigma \in S_\infty$, $\sigma(\xi_{i_1}^{r_1}..) = \xi_{\sigma(i_1)}^{r_1}\ldots$. Given now $\pi = 1^{r_1}2^{r_2}\ldots \in \Pi(n)$, consider the orbit of $\xi_1^{r_1}\xi_2^{r_2}$.. under the action. $\langle\pi\rangle$ is defined as the formal sum in Ω of all the elements in this orbit. In old-fashioned terminology, $\langle\pi\rangle$ is obtained by starting with $\xi_1^{r_1}\xi_2^{r_2}\ldots$ and then symmetrizing.

$\langle\pi\rangle$ is symmetric and homogeneous of degree n in the ξ's, hence by the fundamental theorem of symmetric functions, a polynomial in the a_n's, and indeed an integral combination of the a_π, $\pi \vdash n$. On the other hand, every symmetric function, and in particular every a_π, is clearly writeable in terms of monomial symmetric functions. (Note, for instance, that $a_n = \langle 1^n\rangle$, and $h_n = \sum_{\pi \vdash n} \langle\pi\rangle$). Thus the set of monomial symmetric functions $\{ \langle\pi\rangle \mid \pi \vdash n\}$ forms a basis of Λ_n.

Depending on the problem encountered, one or another basis of Λ_n may be easiest to deal with. For example, when multiplying in Λ, the a_π's or h_π's are most convenient: $a_{\pi_1}.a_{\pi_2} = a_{\pi_1 \cdot \pi_2}$ where $\pi_1 \cdot \pi_2$ is the partition obtained by adjoining π_1 and π_2. (I.e., if $\pi_1 = (1^{r_1}2^{r_2}..)$, $\pi_2 = (1^{s_1}2^{s_2}..)$, then $\pi_1 \cdot \pi_2 = (1^{r_1+s_1}2^{r_2+s_2}..)$.) The multiplication of monomial symmetric functions is on the other hand more reminiscent of classical combinatorics. For example, consider the product $\langle 1^3 3\rangle . \langle 1^2 2\rangle$. The factors are of weight 6 and 4 respectively, so the product is of weight 10. Hence

$$\langle 1^3 3\rangle . \langle 1^2 2\rangle = \sum_{\pi \vdash 10} a_\pi \langle\pi\rangle \quad \text{for some integers } a_\pi .$$

Let us find a_{π_0} where $\pi_0=(1^2 2^2 4)$. a_{π_0} occurs as the coefficient of $\xi_1\xi_2\xi_3^2\xi_4^2\xi_5^4$ and can be interpreted as the number of times this particular term arises as the product of terms of the forms $\xi_{i_1}\xi_{i_2}\xi_{i_3}\xi_{i_4}^3$ $(i_1< i_2< i_3,\ i_4\neq i_1,i_2,i_3)$ and $\xi_{j_1}\xi_{j_2}\xi_{j_3}^2$ $(j_1< j_2,\ j_3\neq j_1,j_2)$. We first observe that the term $\xi_{i_4}^3$ can only contribute to ξ_5^4 so $i_4=5$ and since ξ_5^4 occurs in the product but only ξ_5^3 in the first term, either ξ_{j_1} or ξ_{j_2} of the second term must be ξ_5. Since $j_2>j_1$, & no ξ_j, $j>5$ occurs in the result, it must be that $\xi_{j_2}=\xi_5$. Thus the product must be of the form

$$\xi_{i_1}\xi_{i_2}\xi_{i_3}\xi_5^3 \cdot \xi_{j_1}\xi_5\xi_{j_3}^2 = \xi_1\xi_2\xi_3^2\xi_4^2\xi_5^4 .$$

(Here $i_1< i_2< i_3$; $j_1\neq j_3$). Similar argument shows that $j_3=3$ or 4. Having made that choice, j_1 is determined. Etc. Thus the only way that $\xi_1\xi_2\xi_3^2\xi_4^2\xi_5^4$ arises is either as $(\xi_1\xi_2\xi_3\xi_5^3).(\xi_3\xi_5\xi_4^2)$ or as $(\xi_1\xi_2\xi_4\xi_5^3).(\xi_4\xi_5\xi_3^2)$. Hence $a_{(1^2 2^2 4)}=2$.

Similarly $a_{(1^6 2^2)}=0$, $a_{(1^3 2^2 3)}=2$, and $a_{(1^5 23)}=10$. (The eager reader is invited to work out some of these products, and also to give for Λ_3, say, the transformations between the various bases.)

The next set of symmetric functions is not a real basis at all but rather a "rational basis", i.e., a basis of $\Lambda \otimes \mathbb{Q}$ as a vector space over the rational numbers $\mathbb{Q}$. These functions are the <u>power sums</u>. We define

$$s_1 = \xi_1 + \xi_2 + \cdot\cdot \quad = a_1 \quad = \langle 1 \rangle$$

$$s_2 = \xi_1^2 + \xi_2^2 + \cdot\cdot \quad = a_1^2 - 2a_2 \quad = \langle 2 \rangle$$

$$s_3 = \xi_1^3 + \xi_2^3 + \cdot\cdot \quad = a_1^3 - 3a_1a_2 + 3a_a = \langle 3 \rangle$$

$$\vdots$$

Letting $\lambda_t = \sum a_n t^n = \prod (1+\xi_i t)$, we compute Waring's formula:

$$\frac{\lambda_t'}{\lambda_t} = \frac{d}{dt} \log \lambda_t = \frac{d}{dt} \sum \log(1+\xi_i t) = \sum_i \frac{\xi_i}{1+\xi_i t}$$

$$= \sum \xi_i - \sum \xi_i^2 t + \sum \xi_i^3 t^2 - \quad = \sum_{n=0}^{\infty} (-1)^n s_{n+1} t^n .$$

(In terms of $\frac{1}{\lambda_{-t}} = \sum h_n t^n$, $\frac{d}{dt} \log(\frac{1}{\lambda_{-t}}) = \sum_{n=0}^{\infty} s_{n+1} t^n$.)

We can solve this for the a_n's in terms of the s_n's as follows:

$$\frac{\lambda_t'}{\lambda_t} = \sum_{n=0}^{\infty} (-1)^n s_{n+1} t^n$$

Hence

$$\sum_{n=0}^{\infty} (n+1) a_{n+1} t^n = \left(\sum_{n=0}^{\infty} (-1)^n s_{n+1} t^n \right) \left(\sum_{n=0}^{\infty} a_n t^n \right)$$

Equating coefficients of t^n gives <u>Newton's Formulas</u>

$$(-1)^n s_{n+1} + (-1)^{n-1} s_n a_1 + (-1)^{n-2} s_{n-1} a_2 + \cdot\cdot + s_1 a_n = (n+1) a_{n+1}$$

Thus

$$\begin{aligned}
a_1 &= s_1 \\
s_1 a_1 - 2a_2 &= s_2 \\
s_2 a_1 - s_1 a_2 + 3a_3 &= s_3 \\
s_3 a_1 - s_2 a_2 + s_1 a_3 - 4a_4 &= s_4 \\
&\vdots
\end{aligned}$$

Solving for a_n by Cramer's rule (here we step out of Λ_n into $\Lambda_n \otimes Q$ and allow ourselves to divide by integers) we have

$$a_n = \frac{\det \begin{vmatrix} 1 & 0 & \cdot\cdot & & s_1 \\ s_1 & -2 & 0 & \cdot\cdot & s_2 \\ s_2 & -s_1 & 3 & 0 & s_3 \\ \vdots & & & & \\ s_{n-1} & -s_{n-2} & & & s_n \end{vmatrix}}{\det \begin{vmatrix} 1 & 0 & & 0 \\ s_1 & -2 & 0 & 0 \\ s_2 & -s_1 & 3 & 0 \\ \vdots & & & \\ s_{n-1} & \cdot & \cdot & \cdot\ (-1)^{n+1} n \end{vmatrix}}$$

$$= \frac{1}{n!} \det \begin{vmatrix} s_1 & 1 & 0 & 0 & \cdot\cdot & 0 \\ s_2 & s_1 & 2 & 0 & \cdot\cdot & 0 \\ s_3 & s_2 & s_1 & 3 & \cdot\cdot & 0 \\ \vdots & & & & & \\ s_n & s_{n-1} & \cdot & \cdot & \cdot & s_1 \end{vmatrix}$$

(where the last step follows by performing elementary column operations on the numerator, evaluating the denominator, and observing that all the minus signs cancel.)

Given a partition $\pi = (1^{r_1} 2^{r_2} \ldots)$ of an integer n, we write $s_\pi = s_1^{r_1} s_2^{r_2} \ldots$. Then the above expression shows that, modulo division by n!, $\{s_\pi \mid \pi \vdash n\}$ is a basis of Λ_n.

In Chapter Three, a number of identities will be required involving, for each n, a certain element of $\Lambda_n \otimes \Lambda_n$. We will verify these identities here.

$\Lambda \otimes \Lambda$ is contained in $\Omega \otimes \Omega$. The latter has a polynomial basis consisting of 1, and all expressions $\xi_i \otimes 1$, and $1 \otimes \xi_j$, i,j=1,2,3,.. . An element in $\Lambda \otimes \Lambda$ is thus a polynomial $f(\xi_1 \otimes 1, \xi_2 \otimes 1, \ldots; 1 \otimes \xi_1, \ldots)$ symmetrical in each set of variables. It is convenient to write $\xi_i \otimes 1 = x_i$, $1 \otimes \xi_j = y_j$, $x = x_1 + x_2 + \ldots$, and $y = y_1 + y_2 + \ldots$ so that $\lambda_t(x) = \prod(1+x_i t)$, $\lambda_t(y) = \prod(1+y_j t)$. Then f can be expressed as f(x,y). Thus if $\{r_\pi \mid \pi \vdash n\}$ and $\{s_\pi \mid \pi \vdash n\}$ are two bases of Λ_n, $\{r_{\pi_1}(x) s_{\pi_2}(y) \mid \pi_1, \pi_2 \vdash n\}$ is a basis of $\Lambda_n \otimes \Lambda_n$.

Consider the power series $\displaystyle\sum_{n=0}^{\infty} h_n(xy) t^n = \frac{1}{\prod_{i,j}(1-x_i y_j t)}$.

Calculating this product, we get

$$\prod_i \left(\frac{1}{\prod_j (1-x_i y_j t)} \right) = \prod_i \left(\sum_{n=0}^{\infty} h_n(y)(x_i t)^n \right)$$

$$= \sum_{n=0}^{\infty} \left(\sum_{\pi \vdash n} h_\pi(y) \langle\pi\rangle(x) \right) t^n \quad .$$

Thus $h_n(xy) = \sum_{\pi \vdash n} \left(h_\pi(y) \langle\pi\rangle(x) \right) \in \Lambda_n \otimes \Lambda_n$.

By symmetry, $h_n(xy) = \sum_{\pi \vdash n} \left(\langle\pi\rangle(y) h_\pi(x) \right)$.

Another expression for $h_n(xy)$ arises from the identity

$$\frac{d}{dt} \log \frac{1}{\lambda_{-t}(x)} = \sum_{n=0}^{\infty} s_{n+1}(x) t^n \quad .$$

Hence

$$\log \frac{1}{\prod_i (1-x_i t)} = \sum_{n=1}^{\infty} \frac{s_n(x)\; t^n}{n}$$

so
$$\log \frac{1}{\prod_{i,j} (1-x_i y_j t)} = \sum_j \sum_{n=1}^{\infty} \frac{s_n(x)(y_j t)^n}{n} = \sum_{n=1}^{\infty} \frac{s_n(x) s_n(y) t^n}{n}$$

Hence
$$\frac{1}{\prod_{i,j} (1-x_i y_j t)} = \sum h_n(xy) t^n = e^{\left(\sum_{n=1}^{\infty} \frac{s_n(x) s_n(y) t^n}{n} \right)}$$

Calculating the exponential, we get

$$e^{\left(s_1(x)s_1(y)t + s_2(x)s_2(y)\frac{t^2}{2} + s_3(x)s_3(y)\frac{t^3}{3} + \cdots\right)}$$

$$= \sum_{n=1}^{\infty}\left(\sum \frac{s_1^{\alpha}(x)s_1^{\alpha}(y)}{1^{\alpha}\ a!} \quad \frac{s_2^{\beta}(x)s_2^{\beta}(y)}{2^{\beta}\ \beta!} \ \cdots \ \frac{s_n^{\gamma}(x)s_n^{\gamma}(y)}{n^{\gamma}\ \gamma!}\right) t^n$$

where the inside sum is over all sequences $(\alpha,\beta,..,\gamma)$ with $\alpha+2\beta+\ldots+n\gamma = n$.

Given a partition $\pi = (1^{\alpha}2^{\beta}..n^{\gamma})$ we define

$$|\pi| = \frac{n!}{1^{\alpha}\alpha!2^{\beta}\beta!\ldots n^{\gamma}\gamma!}$$

With this notation, the expression inside the parentheses above becomes

$$\sum_{\pi \vdash n} \frac{|\pi|}{n!} \quad s_{\pi}(x)s_{\pi}(y)$$

Thus

$$h_n(xy) = \sum_{\pi \vdash n} \frac{|\pi|}{n!} \ s_{\pi}(x)s_{\pi}(y)$$

Finally, we need an expression due originally to Cauchy. Here we take $x = x_1+\ldots+x_k$, $y = y_1 +\ldots+ y_k$ and work with symmetric functions of k variables. Define $\Delta(x) = \prod_{i<j}(x_i-x_j)$ and $\Delta(y) = \prod_{i<j}(y_i-y_j)$.

<u>Lemma:</u> $$\frac{1}{\prod_{i,j}(1-x_iy_j)} = \frac{1}{\Delta(x)\Delta(y)} \cdot \det\left|\frac{1}{1-x_iy_j}\right|$$

Proof: Consider the product

$$\prod_{i,j}(1-x_iy_j) \cdot \det\left|\frac{1}{1-x_iy_j}\right|$$

Multiply the ith row of the matrix $\left(\frac{1}{1-x_iy_j}\right)$ by the product $\prod_j (1-x_iy_j)$, for each $i = 1,\ldots,k$. The result is

$$\det \begin{vmatrix} \prod_{j\neq 1}(1-x_1y_j) & \prod_{j\neq 2}(1-x_1y_j) & \cdot\cdot \\ \prod_{j\neq 1}(1-x_2y_j) & \prod_{j\neq 2}(1-x_2y_j) & \cdots \\ \cdots & & \cdot\cdot \end{vmatrix} .$$

Think of each entry $\prod_{j\neq j_0}(1-x_iy_j)$ as a polynomial $f_{j_0}(x_i)$ in the variables x_i, with coefficients in the ring of polynomials $\mathbb{Z}[y_1,\ldots,y_k]$. Observe $f_{j_0}(x)$ is of degree k-1 in x. Thus the product is

$$\det \begin{vmatrix} f_1(x_1) & f_2(x_1) & \cdot\cdot \\ f_1(x_2) & f_2(x_2) & \cdot\cdot \\ \cdots & & \cdot\cdot \end{vmatrix}$$

and $f_1(x) = Ax^{k-1}$ + (lower powers of x), $f_2(x) = Bx^{k-2}$ + (lower powers of x) with A and $B \in \mathbb{Z}[y_1,\ldots,y_k]$. Replacing the second column by $-Bf_1+Af_2$ changes the value of the determinant involved only by a factor of A, which does not involve the x's. But it reduces the degree in x of the polynomials in the second column. Similar column operations reduce the 3d column to degree k-3 in x, without changing the degree of the determinant in x. Etc. Thus the original determinant is of degree at most $(k-1)+(k-2)+(k-3)+\ldots+(1)+(0) = \frac{k(k-1)}{2}$ in the x's. On the other hand, the original determinant is an alternating function of the x_i's (i.e., it changes sign if x_i and x_j are switched, for $i\neq j$). Hence it is divisible by $\prod_{i<j}(x_i-x_j) = \Delta(x)$. Since $\Delta(x)$ is also of degree $\frac{k(k-1)}{2}$ in the x's, the result $\frac{1}{\Delta(x)}\prod(1-x_iy_j)\det\left|\frac{1}{1-x_iy_j}\right|$ is independent of the x's. Similarly, the same product divided by $\Delta(y)$ is independent of the y's. Hence

$$\frac{1}{\Delta(x)\Delta(y)}\cdot\prod_{i,j}(1-x_iy_j)\cdot\det\left|\frac{1}{1-x_iy_j}\right|$$

is independent of the x's and y's, so equal to a constant $\varkappa$. To evaluate the constant, consider the equation written in the form

$$\det\left|\prod_{q\neq j}(1-x_iy_q)\right| = \varkappa\cdot\Delta(x)\cdot\Delta(y) \quad .$$

To find the value of $\varkappa$, just evaluate both sides setting $x_i=\frac{1}{y_i}$, $i=1,\ldots,k$. A simple computation shows that $\varkappa=1$. ∎

We now write the above lemma, substituting $yt = y_1t + y_2t + .. Y_kt$ for $y = y_1 + y_2 + \ldots + y_k$, and calculate

$$\frac{1}{\prod(1-x_iy_jt)} = \frac{1}{\Delta(x)\Delta(y)t^{\frac{k(k-1)}{2}}} \det \left| \sum_{n=0}^{\infty} x_i^n y_j^n t^n \right|$$

$$= \frac{1}{\Delta(x)\Delta(y)} \frac{1}{t^{\frac{k(k-1)}{2}}} \sum_{\sigma \in S_k} \operatorname{sgn}(\sigma) \Big(\sum_{n_1} x_{\sigma(1)}^{n_1} y_1^{n_1} t^{n_1} \Big) \Big(\ldots$$

$$\ldots \Big) \Big(\sum_{n_k} x_{\sigma(k)}^{n_k} y_k^{n_k} t^{n_k} \Big)$$

$$= \sum_{n=0}^{\infty} \frac{1}{\Delta(x)\Delta(y)t^{\frac{k(k-1)}{2}}} t^n \Big(\sum \operatorname{sgn}(\sigma) x_{\sigma(1)}^{n_1} x_{\sigma(2)}^{n_2} \cdots x_{\sigma(k)}^{n_k} y_1^{n_1} \cdots y_k^{n_k} \Big)$$

where the sum is over all sums $n_1 + n_2 + .. + n_k = n$ of non-negative integers and all permutations $\sigma \in S_k$ of the x's.

Note that since the terms in the determinant involve $\operatorname{sgn}(\sigma)$, they are alternating, so in any term where $n_i = n_j$, with $i \neq j$, the coefficient is zero. Hence the sum can be taken over all $\sigma \in S_k$ and all sums $n_1 + n_2 + \ldots + n_k = n$, $n_i \neq n_j$ for $i \neq j$. Hence we can sum over those expressions $n = n_1 + n_2 + \ldots + n_k$, with the n_i in decreasing order, by adding in the extra summation over all permutations of the y's. The expression in the parentheses becomes

$$\operatorname{sgn}(\sigma\tau^{-1}) x_{\sigma(1)} x_{\sigma(2)} \cdots x_{\sigma(k)} y_{\tau(1)} y_{\tau(2)} \cdots y_{\tau(k)}$$

with the sum being taken over all sequences $n_1 > n_2 > n_3 > .. > n_k > 0$, with $n_1 + \ldots + n_k = n$, and all $\sigma, \tau \in S_k$.

Finally, define $\lambda_k=n_k$, $\lambda_{k-1}=n_{k-1}-1$, . . ., $\lambda_1=n_1+(k-1)$ so that $\lambda_1\geq\lambda_2\geq\ldots\geq\lambda_k$ and $\lambda_1+\lambda_2+\ldots+\lambda_k=n-\frac{k(k-1)}{2}$. The whole expression $\frac{1}{\prod(1-x_iy_jt)}$ then becomes equal to

$$\sum_{n=0}^{\infty}\frac{t^{n-\frac{k(k-1)}{2}}}{\Delta(x)\Delta(y)}\left(\sum_{\substack{\lambda_1\geq\lambda_2\geq\ldots\geq\lambda_k\geq 0\\ \lambda_1+\lambda_2+\ldots+\lambda_k=n-\frac{k(k-1)}{2}\\ \sigma,\tau\in S_k}}\operatorname{sgn}(\sigma)\operatorname{sgn}(\tau)x_{\sigma(1)}^{\lambda_1+k-1}\cdots x_{\sigma(k)}^{\lambda_k}y_{\tau(1)}^{\lambda_1+k-1}\cdots y_{\tau(k)}^{\lambda_k}\right)$$

We define, for a partition $\lambda=(\lambda_1,\lambda_2,\ldots,\lambda_k)$ the <u>Schur function</u> $\{\lambda\}$ by

$$\{\lambda\}=\frac{\sum_{\sigma\in S_k}\operatorname{sgn}(\sigma)\,x_{\sigma(1)}^{\lambda_1+k-1}\,x_{\sigma(2)}^{\lambda_2+k-2}\cdots x_{\sigma(k)}^{\lambda_k}}{\Delta(x)}$$

Comparing coefficients of $m=n-\frac{k(k-1)}{2}$ in the identity above, we get

$$h_m(xy)=\sum_{\lambda\vdash m}\{\lambda\}(x)\{\lambda\}(y) .$$

Notice that $\{\lambda\}$ is given as a symmetric function of k variables $x_1,x_2,\ldots,x_k$. (The symmetry follows from the fact that both $\Delta(x)$ and $\sum_{\sigma}\operatorname{sgn}(\sigma)x_{\sigma(1)}^{\lambda_1+k-1}\ldots x_{\sigma(k)}^{\lambda_k}$ are alternating functions of $x_1,x_2,\ldots,x_k$, and so their ratio is unchanged by any permutation of the x_i's. Hence $\{\lambda\}$ can be expressed as a polynomial in the elementary symmetric functions of the x_i's:

$$x = x_1+x_2+..+x_k, \ \lambda^2 x = x_1x_2+x_1x_3+x_2x_3+...+x_{k-1}x_k, \ . \ . \ .$$

Thus $\{\lambda\}(x) = F_\lambda(x, \lambda^2 x, \ldots, \lambda^k x)$ for some polynomial F with integer coefficients. Now, a priori, F_λ could depend on the original integer k chosen (k = the number of x_i's), so we should write $F_{\lambda,k}$. But it will be a corollary of the interpretation of these functions in Chapter III that if $k \geq n$, and $\lambda \vdash n$, then $F_{\lambda,k} = F_{\lambda,k+1}$. (The reader is invited to prove this directly.) Hence F_λ is independent of k, even though to calculate it for any particular $\lambda \vdash n$, a particular number $k \geq n$ must be chosen and fixed.

Exercise: $\{1^n\} = a_n$, $\{n\} = h_n$.

Schur functions, also called S-functions in the literature, have been defined and studied in a number of ways, not just in terms of representation theory as we shall do in Chapter III. For other approaches, see Littlewood ([29]), Read ([34]), and Stanley ([43]).

The Schur functions can, of course, be expressed in terms of monomial symmetric functions. Straightforward but tedious calculation shows:

$$\{1\} = \langle 1\rangle$$

$$\begin{aligned}\{2\} &= \langle 2\rangle + \langle 11\rangle\\ \{11\} &= \langle 11\rangle\end{aligned}$$

$$\begin{aligned}\{3\} &= \langle 3\rangle + \langle 21\rangle + \langle 111\rangle\\ \{21\} &= \langle 21\rangle + 2\langle 111\rangle\\ \{111\} &= \langle 111\rangle\end{aligned}$$

$$\begin{aligned}\{4\} &= \langle 4\rangle + \langle 31\rangle + \langle 22\rangle + \langle 211\rangle + \langle 1111\rangle\\ \{31\} &= \langle 31\rangle + \langle 22\rangle + 2\langle 211\rangle + 3\langle 1111\rangle\\ \{22\} &= \langle 22\rangle + \langle 211\rangle + 2\langle 1111\rangle\\ \{211\} &= \langle 211\rangle + 3\langle 1111\rangle\\ \{1111\} &= \langle 1111\rangle\end{aligned}$$

In Chapter Three, we will give a more natural definition of these functions, related to the representation theory of the symmetric groups.

4. Adams Operations

Let Λ be the free λ-ring on one generator and α be an element of Λ. Regarding Λ as the ring of symmetric polynomials in an infinite number of variables $\xi_1, \xi_2, \ldots$, α might be described, say, by giving its description as a symmetric function. Thus, since $\Lambda = \mathbf{Z}[1, a_1, a_2, \ldots]$, where a_i is the ith elementary symmetric function of the ξ's, α is uniquely expressible as a polynomial in the a_i: $\alpha = F(a_1, a_2, \ldots)$. Given a λ-ring R, α becomes a natural operation on R by taking, for each $r \in R$, $\alpha(r) = F(r, \lambda^2 r, \lambda^3 r, \ldots)$.

In this section we will investigate the operations associated with several types of symmetric functions, most notably the power sum functions $s_n = \sum (\xi_i)^n$ which give rise to the Adams operations. (One exception - the operations associated with the Schur functions is postponed until Chapter III.)

These operations can most easily be described (and hence named) in the context of the example of the λ-ring R constructed from finite dimensional vector spaces over a fixed field F (p. 5).

The elementary symmetric functions give rise, as per design, to the exterior powers: for $V \in R$, we wrote $a_n(V) = \Lambda^n V$. The homogeneous power sum h_n gives the <u>n-fold symmetric power</u> $h_n(V) = \mathrm{Symm}^n V$ for $V \in R$. Just as with exterior powers there is an additivity formula

$$h_n(V_1+V_2) = \sum_{i=0}^{n} h_i(V_1)h_{n-i}(V_2)$$

To see this, write $h_t(V) = \sum h_n(V)t^n$. Then the definition of h_n gives an identity $h_t(V)\lambda_{-t}(V)=1$, so the additivity formula for h_n follows from that for λ^n.

The monomial symmetric functions $\langle\pi\rangle$ provide operations, also denoted $\langle\pi\rangle$. In this case there is an addition formula

$$\langle\pi\rangle\ (V_1+V_2) = \sum_{\pi_1\pi_2=\pi} \langle\pi_1\rangle(V_1)\langle\pi_2\rangle(V_2)$$

To prove this, it is enough, using the Verification Principle, to assume V_1 and V_2 are sums of 1st degree elements. The argument can then be carried out with the explicit definition of $\langle\pi\rangle$ and the $\langle\pi_i\rangle$'s, just as in the additivity formula for the exterior powers (p.6). Details are left to the reader. (Or see [32], p.91, Theorem 33).

The main subject of this section is the set of operations associated with the power sums s_n. We define the <u>n^{th} Adams Operation</u>, Ψ^n, by $\Psi^n(V) = s_n(V)$, for V in a λ-ring R.

<u>Proposition</u>: Let a,b be elements in a λ-ring R, and n,m integers ≥ 1

Then

1) $\Psi^1(a) = a$

2) $\Psi^n(1) = 1$

3) $\Psi^n(a+b) = \Psi^n(a) + \Psi^n(b)$

4) $\Psi^n(ab) = \Psi^n(a)\Psi^n(b)$

5) $\Psi^n(\lambda^m(a)) = \lambda^m(\Psi^n(a))$

6) $\Psi^n(\Psi^m(a)) = \Psi^{nm}(a) = \Psi^m(\Psi^n(a))$

Thus each Ψ^n is a λ-ring endomorphism of R, and the map $n \rightsquigarrow \Psi^n$ is a ring homomorphism $\mathbb{Z} \longrightarrow \text{End } R$.

<u>Proof</u>: By the Verification Principle, we can assume that a and b are each sums of elements of degree 1. Then using the original definition of the power sums s_n (p.35) all these are clear. ∎

The Adams operations also serve to distinguish binomial elements of λ-rings:

<u>Proposition</u>: Let R be a λ-ring and $a \in R$. Then a is binomial iff $\Psi^n(a) = a$ for all $n \geq 1$.

<u>Proof</u>: Use the identity $\frac{d}{dt}\log(\lambda_t(a)) = \sum_{n=0}^{\infty} (-1)^n \Psi^{n+1}(a)t^n$. Then $\lambda_t(a) = (1+t)^a$ iff the right hand side is $\sum (-1)^n a t^n = \frac{a}{1+t}$. ∎

<u>Corollary</u>: For all $m \in \mathbb{Z}$, $\Psi^n(m) = m$. ∎

Hence, given any λ-ring R, we can pick out a maximal binomial sub-λ-ring R_1 of R by $R_1 = \{x \mid \Psi^n(x)=x, \text{ all } n \geq 1\}$. This subring will at least include the unit 1, so will include a copy of $\mathbb{Z}$.

The first proposition above has a converse, which will be useful in verifying that various pre-λ-rings are in fact λ-rings. First we need a definition: a ring R is <u>torsion-free</u> if for any nonzero element $r \in R$, and any integer $n \geq 1$, $nr = r+\ldots+r$ (n summands) is also nonzero.

<u>Theorem</u>: Let R be a torsion-free pre-λ-ring. Let operations Ψ^n be defined by, for $a \in R$, $\frac{d}{dt}\log\lambda_t(a) = \sum_{n=0}^{\infty}(-1)^n\Psi^{n+1}(a)t^n$.

(So in particular, we have $\Psi^1(a)=a$, $\Psi^n(a+b)=\Psi^n(a)+\Psi^n(b)$, all a,b,n.) Suppose $\Psi^n(1)=1$, $\Psi^n(ab)=\Psi^n(a)\Psi^n(b)$, and $\Psi^n(\Psi^m(a))=\Psi^{nm}(a)$ for all $a,b \in R$ and integers n,m. Then R is a λ-ring.

To prove the theorem, we make a general definition: A <u>pre-Ψ-ring</u> R is a commutative ring R with unit, together with a set of operations $\Psi^n: R \longrightarrow R$, $n \geq 1$, satisfying $\Psi^1(a)=a$, and $\Psi^n(a+b)=\Psi^n(a)+\Psi^n(b)$, for all $a,b \in R$, and integers n. A <u>Ψ-ring</u> is a pre-Ψ-ring also satisfying $\Psi^n(1)=1$, $\Psi^n(ab)=\Psi^n(a)\Psi^n(b)$, and $\Psi^n(\Psi^m(a))=\Psi^{nm}(a)$ for all $a,b \in R$ and integers $n,m \geq 1$.

Given any commutative ring R with unit, let R^{ω} be the set of countable sequences $\{(r_1,r_2,r_3,..) \mid r_i \in R\}$. R^{ω} is given a ring structure by defining addition and multiplication coordinatewise. For each integer $n \geq 1$, we define an operation $\Psi^n: R^{\omega} \longrightarrow R^{\omega}$, by $\Psi^n((r_1,r_2,\ldots)) = (r_n,r_{2n},r_{3n},\ldots)$.

<u>Proposition:</u> R^{ω} is a Ψ-ring. If R is a pre-Ψ-ring, then R is a Ψ-ring if and only if the map $\Psi: R \longrightarrow R^{\omega}$, defined by $\Psi(r) = (\Psi^1(r), \Psi^2(r), \ldots)$, is a homomorphism of Ψ-rings.

<u>Proof</u>: Clear.

Note that since Ψ^1 is the identity map, the map $\Psi: R \longrightarrow R^{\omega}$ is one-one.

Let R be a torsion-free pre-λ-ring. Suppose there are operations $\lambda^n: R \longrightarrow R$, $n \geq 0$, so that $\frac{d}{dt} \log\lambda_t(x) = \sum_{n=1}^{\infty} (-1)^n \Psi^{n+1}(x) t^n$, all $x \in R$.

E.g., following p. , we can calculate

$$n!\, \lambda^n(x) = \det \begin{vmatrix} \Psi^1(x) & 1 & 0 & \cdot & \cdot \\ \Psi^2(x) & \Psi^1(x) & 2 & \cdot & \cdot \\ & & & & \\ \Psi^n(x) & & & & \Psi^1(x) \end{vmatrix}$$

Then we suppose that for each $n > 1$, and each $x \in R$ the element defined by the determinant on the right hand side is divisible by $n!$ in R, so that $\lambda^n(x)$ is defined. (The torsion-free provision guarentees that this division is well-defined if defined.) Note that if

R contains a field of characteristic zero, division is always possible so the λ^n's automatically exist.

As usual, the power series $\lambda_t = \Sigma \lambda^n t^n$ gives a map $R \longrightarrow 1+R[[t]]^+$ and we can define a map L making the following diagram commute:

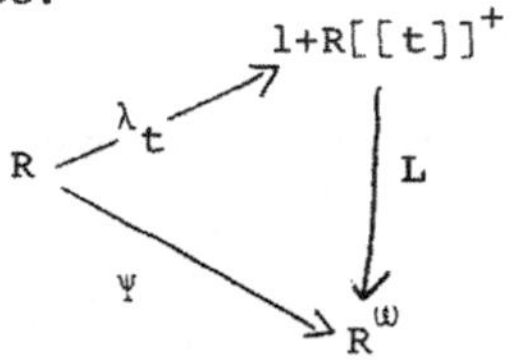

L is defined by: $L(1+a_1t+a_2t^2+\ldots) = (r_1,r_2,\ldots)$ if
$\frac{d}{dt}\log(1+a_1t+a_2t^2+\ldots) = (-1)^n r_{n+1}t^n.$

<u>Proposition:</u> 1) $L(x"+"y) = L(x) + L(y)$

2) $L("1") = 1$

3) $L(x"\cdot"y) = L(x)L(y)$

4) $L("\Psi^n"(x)) = \Psi^n L(x)$

5) If R is torsion-free, L is one-one.

6) If R contains a field of characteristic zero, L is onto, and hence an isomorphism of Ψ-rings.

<u>Proof</u>: (Recall first that the "..." notation refers to the λ-ring operations in $1+R[[t]]^+$ - thus "Ψ^n" refers to the Adams operations there.) 1) and 2) are easy. Then to check 3),4) it is sufficient to take x and y to be of degree 1, giving again an easy verification. For 5), we can use 1), so the injectivity of L follows from the easy observation that the kernel of L is trivial. For 6) the

inverse map $L^{-1}: R^{\omega} \longrightarrow 1+R[[t]]^{+}$ is easily calculated as $L^{-1}((b_1,b_2,\ldots)) = \exp(-g(t))$, where $g(t)=\sum_{n+1}^{\infty}(-1)^n b_n t^n$. ∎

Corollary: If R is an algebra over a field k of characteristic zero (e.g., $k=\mathbb{Q}$), then so is $1+R[[t]]^{+}$.

Proof: R^{ω} is certainly a k-algebra, and under the hypothesis, is isomorphic to $1+R[[t]]^{+}$. ∎

The Proof of the original theorem is now accomplished. If R has λ-operations, and hence Ψ-operations, and is torsion-free, R is a λ-ring iff R is a Ψ-ring. A useful restatement of the theorem is the following proposition.

Proposition: Let R be a torsion-free ring, and S be any ring. Suppose there is given a map of sets $\varphi : S \longrightarrow 1+R[[t]]^{+}$. Then φ is a ring homomorphism iff the composite map $S \xrightarrow{\varphi} 1+R[[t]]^{+} \xrightarrow{L} R^{\omega}$ is a ring homomorphism. If S is a pre-λ-ring, φ preserves the λ-structure iff the composite map $L\varphi$ does. ∎

There is an interesting application of this proposition to the algebraic geometry of varieties over finite fields (which the non-geometeer may ignore since it will not be relevant to the sequel.) Let k be a finite field and S the Grothendieck ring of varieties defined over k, where sum comes from the disjoint union of varieties and product from the cartesian product. Let R = $\mathbb{Z}$, and $\zeta : S \to 1+\mathbb{Z}[[t]]^+$ be the assignment to each variety X its zeta-function $\zeta_X(t)$. (See Swinnerton-Dyer [44] for definitions.) Then in the composite map $S \to \mathbb{Z}^\omega$, $L(\zeta_X(t))_n$ is the number of points on X with coordinates in the extension field k' of k, where the degree of k' over k is n. Since the number of rational points on a disjoint union of two varieties is the sum of the rational points on each variety, and similarly for products, the restated proposition implies that ζ is a ring homomorphism $R \to 1+\mathbb{Z}[[t]]^+$.

One might hope that ζ is also a map of λ-rings. Alas, it seems not so. The obvious λ-structure to put on R would be to take the n-fold symmetric power of a variety X to be $h_n(X)$, and define the other operations accordingly. Under this definition affine 1-space A^1 is of degree 1, but its square $A^1 \times A^1 = A^2$ is not, and in a λ-ring, degree is multiplicative. Hence R is not a λ-ring in this definition, and no other definition seems nearly as plausible. We should remark, however, that this category of varieties over k is a good example of a category which has sums, products, and symmetric powers, but whose Grothendieck ring is not a λ-ring.

We now use the theorem to construct a general class of λ-rings. Let S be a set, K a field of characteristic zero, and K(S) the set of all maps from S to K. The sum and product of two maps is defined as usual, making K(S) a commutative ring with identity. Suppose there is given on S a set of maps $\sigma_n: S \longrightarrow S$, $n=1,2,\ldots$, satisfying

i) σ_1 is the identity map

ii) $\sigma_n \sigma_m = \sigma_{nm}$.

We then define operations $\psi^n: K(S) \longrightarrow K(S)$ by, for $f: S \longrightarrow K$, and $s \in S$, $\psi^n(f(s)) = f(\sigma_n(s))$. As in the theorem, since division by n! is possible in K(S) for every integer n, we can define operations λ^n by

$$\lambda^n = \frac{1}{n!} \det \begin{vmatrix} \psi^1 & 1 & 0 & \\ \psi^2 & \psi^1 & 2 & 0 \\ & & & \\ \psi^n & & & \psi^1 \end{vmatrix}$$

and we can conclude that the ψ^n are the Adams operations corresponding to the operations λ^n. By the theorem, since K(S) is clearly a ψ-ring, it is also a λ-ring.

This construction applies in particular to the case when one is given a group G, and S is the set of conjugacy classes of G. σ_n is the map induced on S by the operation of taking the nth power in G. K(S) is then called the <u>λ-ring of central functions on G with values in K</u>.

The map $1+R[[t]]^+ \xrightarrow{L} R^\omega$ classically appears in another context - that of universal Witt rings - which we now describe. Let R be a commutative ring with unit (assumed torsion-free for the moment) and let W_R be the set of all ω-tuples $(w_1, w_2, \ldots)$ of elements of R. As a set, $W_R = R^\omega$, but we will put a rather different ring structure on W_R.

Consider the map $M: W_R \longrightarrow R^\omega$ defined by

$$M((w_1, w_2, \ldots)) = (r_1, r_2, \ldots) \quad \text{where} \quad r_n = \sum_{d|n} d w_d^{n/d}$$

Thus

$$\begin{aligned} r_1 &= w_1 \\ r_2 &= w_1^2 + 2w_2 \\ r_3 &= w_1^3 + 3w_3 \\ r_4 &= w_1^4 + 2w_2^2 + 4w_4 \\ &\text{etc.} \end{aligned}$$

If R is torsion-free, M is a one-one map and we can identify W_R with its image $M(W_R) \subset R^\omega$.

<u>Proposition</u>: $W_R(M)$ is closed under sum and product in R^ω. Indeed there are polynomials F_i, G_j with integer coefficients such that

$$M((w_1, w_2, \ldots)) + M((w_1', w_2', \ldots)) = M((F_1(w_1, w_1'), F_2(w_1, w_2, w_1', w_2'), \ldots))$$

(here F_n depends on two sets of variables: $\{w_i \mid i \text{ divides } n\}$ and $\{w_i' \mid i \text{ divides } n\}$) and

$$M((w_1, w_2, \ldots)) \cdot M((w_1', w_2', \ldots)) = M((G_1(w_1, w_1'), G_2(w_1, w_2, w_1', w_2', \ldots))$$

(where similarly G_n is a function of the two sets of variables).

Hence using these polynomials to define addition and multiplication in W_R, W_R becomes a commutative ring with identity - the

(Universal) ring of Witt vectors of R.

For the proof of the proposition, we must construct the polynomials F_i, G_j, which can be accomplished by just proving the special case for the ring $R = \mathbb{Z}$. For, once we have the polynomials, they define operations in the set W_R, for any ring R, (torsion-free or not). The ring axioms, associativity etc., will follow for an arbitrary ring because they are true over $\mathbb{Z}$, hence are polynomial identities valid in all rings.

Actually, to carry out the proof, we will just assume that R is torsion-free (an obvious property of $\mathbb{Z}$). Then the map $W_R \xrightarrow{M} R^{\omega}$ is injective. Also, the map $1+R[[t]]^+$ is injective. The proposition follows by observing that these two inclusions give the same subset of R^{ω}! Indeed, we can define $f: W_R \longrightarrow 1+R[[t]]^+$ by, for $w = (w_1, w_2, \ldots) \in W_R$,

$$f(w) = \prod_d (1 - w_d(-t)^d) \quad .$$

Then $L \circ f(w) = \frac{d}{dt} \log\left(\prod_d (1 - w_d(-t)^d)\right)$

$$= \sum_d \left(\frac{d\, w_d(-t)^{d-1}}{1-w_d(-t)^d}\right) = \sum_d \left(\frac{-d\, w_d(-t)^d}{t}\right)(1 + w_d(-t)^d + w_d^{\,2}(-t)^{2d} + \ldots$$

$$= \sum_n \frac{1}{t}(-1)\left(\sum_{d|n} d\, w_d^{\,n/d}(-t)^n\right) = \sum_n (-1)^{n+1}\left(\sum_{d|n} d\, w_d^{\,n/d}\right) t^{n-1}$$

$$= \sum_n (-1)^{n-1}\, r_n\, t^{n-1} \quad .$$

Hence, by the definition of the map L, $L\,f(w) = (r_1, r_2, \ldots)$. Since M is one-one, so is f, so (for R torsion-free) W_R can be considered via f as a subset of $1+R[[t]]^+$. To see that W_R is all of $1+R[[t]]^+$, it is necessary only to show that f is an invertible map, a fact which is obvious once we write out the definition of f in more detail:

Given $w = (w_1, w_2, w_3, \ldots) \in W_R$,

$$f(w) = \prod_d (1 - w_d(-t)^d)$$

$$= (1+w_1 t)(1-w_2 t^2)(1+w_3 t^3)(1-w_4 t^4)\ldots$$

$$= 1 + (w_1)t + (-w_2)t^2 + (w_3 - w_1 w_2)t^3 + (-w_4 + w_1 w_3)t^4$$

$$+ (w_5 - w_4 w_1 - w_2 w_3)t^5 + (-w_6 + w_5 w_1 + w_4 w_2 - w_3 w_2 w_1)t^6 + \ldots$$

The coefficient of t^n is $(-1)^n \sum (-1)^k w_{n_1} w_{n_2} \ldots w_{n_k}$ where the sum is over all partitions $n_1 > n_2 > \ldots > n_k$ of n into k distinct parts, and all integers k. Hence if $1+a_1 t + a_2 t^2 + \ldots \in 1+R[[t]]^+$ is to be in the image of f, we must solve the equations

$$a_1 = w_1$$

$$a_2 = -w_2$$

$$a_3 = w_3 - w_1 w_2$$

etc.

Since in the nth equation w_n occurs in only one term with the coefficient ± 1, the equations can be solved inductively for the w's in terms of the a's.

The first few terms are

$w_1 = a_1$

$w_2 = -a_2$

$w_3 = a_3 + a_1a_2$

$w_4 = -a_4 + a_3a_1 + a_2a_1^2$

$w_5 = a_5 - a_4a_1 - a_3a_2 - a_1a_2^2 + a_1^2a_3 + a_1^3a_2$

etc.

Hence W_R is isomorphic as a ring to $1+R[[t]]^+$, for any ring R.

(It should be mentioned here that several authors ([11], [24], [5] have set up the isomorphism $W_R \longrightarrow 1+R[[t]]^+$ as $(w_1, w_2, \ldots) \rightsquigarrow \prod_d (1 - w_d t^d)$. This forces them to take the "multiplication" of first degree elements in $1+R[[t]]^+$ to satisfy $(1-at)(1-bt)=(1-abt)$. Our choice of f is governed by our rule that first degree elements should multiply $(1+at)(1+bt)=(1+abt)$ - so $(1-at)(1-bt)=(1+(-a)(-b)t)=(1+abt)$. Indeed our $\prod_d (1-w_d(-t)^d)$ is to their $\prod_d (1-w_d t^d)$ as the expression $\lambda_t(x)=\sum\lambda^n(x)t^n$ is to $\lambda_{-t}(x) = (\sum h_n(x)t^n)^{-1}$.)

Additional properties of this construction of W_R and its relation to the p-vectors of Witt can be found in the references cited above.

CHAPTER II : THE REPRESENTATION THEORY OF GROUPS

The following notation will be used throughout this chapter:

All groups are finite

All vector spaces are finite-dimensional over the complex numbers.

$\mathbb{C}$ denotes the complex numbers, and for $z\in\mathbb{C}$, z^* is its conjugate.

If V and W are vector spaces, Hom(V,W) is the vector space of linear maps from V to W. Aut V is the subset of Hom(V,V) consisting of invertible elements. If dimV = n, picking a basis of V gives an isomorphism of groups AutV $\cong$ Gl(n,$\mathbb{C}$), the group of invertible $n\times n$ matrices over $\mathbb{C}$.

Tensor products, $\otimes$, are over $\mathbb{C}$ unless otherwise specified.

We take Aut V to act on V by left multiplication ($f(g(v))=(fg)(v)$) so the elements of V are "column vectors" rather than "row vectors".

S_n denotes the symmetric group of permutations of n objects. Denote these objects 1,2,3,...,n. Let $\sigma\in S_n$ be the transposition $\sigma(1)=2$, $\sigma(2)=1$, $\sigma(j)=j$, $j>2$. Let τ be the n-cycle $\tau(i)=i+1$, $i=1,2,..,n-1$ and $\tau(n)=1$. Then S_n is generated by σ, τ subject to the relations $\sigma^2=1$, $\tau^n = 1$, and $\sigma\tau = \tau^{n-1}\sigma$. In particular, S_3 consists of the elements $1,\tau,\tau^2,\sigma,\sigma\tau,\sigma\tau^2$ which we usually take in this order.

1. The Representation Ring of a Finite Group

A <u>representation</u> of a group G, of <u>degree n</u>, is a homomorphism $\rho : G \to Gl(n, C)$.

For example, if $G = S_3$, we can assign

$$\rho(1) = \begin{pmatrix} 1 & 0 \\ 0 & 1 \end{pmatrix} \qquad \rho(\sigma) = \begin{pmatrix} 0 & 1 \\ 1 & 0 \end{pmatrix}$$

$$\rho(\tau) = \begin{pmatrix} 0 & -1 \\ 1 & -1 \end{pmatrix} \qquad \rho(\sigma\tau) = \begin{pmatrix} 1 & -1 \\ 0 & -1 \end{pmatrix}$$

$$\rho(\tau^2) = \begin{pmatrix} -1 & 1 \\ -1 & 0 \end{pmatrix} \qquad \rho(\sigma\tau^2) = \begin{pmatrix} -1 & 0 \\ -1 & 1 \end{pmatrix}$$

Thus for $x, y \in S_3$, the matrix product of $\rho(x)$ and $\rho(y)$ is equal to $\rho(xy)$.

While this is a nice concrete definition, we prefer to take a coordinate-free approach to the problem.

Let V be a vector space, and G a group. A <u>representation of G in V</u> is a homomorphism $\rho : G \to \text{Aut } V$. The <u>degree</u> of the representation is the dimension of V. (Of course, picking a basis of V gives an isomorphism of Aut V with $Gl(n, \mathbb{C})$, so this does generalize the concrete definition above.)

While strictly speaking, the representation of G is given by the pair (V, ρ), we will often speak of "the representation V" if ρ is clear from the context. In a similar way, given $\rho : G \to \text{Aut } V$, and $g \in G$, we should refer to the associated operation on V as $\rho(g)$ and write $\rho(g)(v)$ for $v \in V$. But it is usually simpler to write g both for the element of G and for $\rho(g)$, and to write gv for $\rho(g)(v)$.

In other common terminology, (V, ρ) is also called a <u>linear</u> representation of G (to distinguish it from a permutation representation, defined below) or a <u>complex</u> representation (to distinguish it from the more general notion of a map $\rho: G \to Gl(n,R)$, where R is an arbitrary commutative ring). V is called a (left) <u>G-module</u> and ρ is said to give <u>an action of G on V</u>.

Given an action of G on V, a subspace W of V is <u>invariant</u> under G if $gw \in W$ for all $g \in G$ and $w \in W$. An element $v \in V$ is <u>invariant</u>, or a <u>fixed point</u>, if $gv=v$ for all $g \in G$. We write V^G for the set of invariant elements in V. V^G is a subspace of V and is invariant.

A <u>map of G-modules</u> $f:(V_1, \rho_1) \to (V_2, \rho_2)$ is a linear map of vector spaces $f:V_1 \to V_2$ satisfying, for any $v \in V_1$ and $g \in G$, $f(gv)=gf(v)$. Let $Hom_G(V_1,V_2)$ be the set of maps of G-modules V_1 to V_2.

The set of all linear maps from V_1 to V_2, $Hom(V_1,V_2)$ gives a representation of G by defining, for $\varphi: V_1 \to V_2$ and $g \in G$, $g\varphi: V_1 \to V_2$ is the map $g\varphi(v) = g(\varphi(g^{-1}v))$ for all $v \in V_1$. $Hom_G(V_1,V_2)$ is then exactly $Hom(V_1,V_2)^G$.

A map of G-modules $f(V_1, \rho_1) \longrightarrow (V_2, \rho_2)$ is an <u>isomorphism</u> if there is a map of G-modules $f':(V_2, \rho_2) \longrightarrow (V_1, \rho_1)$ so that the composites ff' and f'f are the identity maps on V_2 and V_1. (V_1, ρ_1) and (V_2, ρ_2) are <u>isomorphic</u> (also called <u>equivalent</u>) if there exists such an isomorphism. Note that, given (V_1, ρ_1) and (V_2, ρ_2), V_1 and V_2 can be isomorphic (indeed identical) as vector spaces without being isomorphic as G-modules.

We are interested really only in isomorphism classes of representations. Given a representation $\rho: G \to \mathrm{Aut}\, V$, each choice of basis for V gives a matrix representation of G, $G \longrightarrow Gl(n, C)$, and all these are isomorphic. But some choices of basis of V are more natural than others, as the following proposition shows.

<u>Proposition</u>: Let $\rho: G \to \mathrm{Aut}\, V$ be a representation of G. Then there is an inner product on V, call it $\langle -,- \rangle$, such that for each $g \in G$, and all v_1, v_2 in V, $\langle v_1, v_2 \rangle = \langle gv_1, gv_2 \rangle$. Hence every representation of G is isomorphic to a representation by unitary matrices.

<u>Proof</u>: Pick any basis $e_1, \ldots e_n$ of V and let $(-,-)$ be the usual inner product with respect to this basis: $(\sum_i a_i e_i, \sum_j b_j e_j) = \sum_i a_i b_i^*$, the asterisk denoting complex conjugation. Now define, for v_1, v_2 in V,

$$\langle v_1, v_2 \rangle = \frac{1}{|G|} \sum_{g \in G} (gv_1, gv_2)$$

This is clearly bilinear, skew-symmetric and nondegenerate, since the usual inner product is. Also, for $h \in G$,

$$\langle hv_1, hv_2 \rangle = \frac{1}{|G|} \sum_{g \in G} (hgv_1, hgv_2)$$

$$= \frac{1}{|G|} \sum_{g \in G} (gv_1, gv_2) = \langle v_1, v_2 \rangle$$

- where the middle equality holds since the set of all elements hg, $g \in G$ is the same as the set of all g, $g \in G$.

Using the Gram-Schmid process, we can find an orthonormal basis of V for the inner product $\langle -,- \rangle$. With respect to this basis, all the matrices assigned to G are unitary. ∎

<u>Corollary</u> Let $\rho : G \to \operatorname{Aut} V$ be a representation of G, and let g be an element of G. Then there is a basis of V in which the matrix of $\rho(g)$ is diagonal.

Proof: Every unitary matrix is diagonalizable. QED. Another way to see this is to observe that every matrix A satisfying $A^n = I$, for some $n \geq 1$, and having entries in a field of characteristic zero, must have a diagonal matrix for its Jordan form. ∎

We will see later that we can simultaneously diagonalize all the matrices of G if and only if G is abelian.

Representations arise in many ways. The trivial representation of degree n of a group G is the assignment of the $n \times n$ unit matrix to each element $g \in G$. For $n=0$, this is the zero representation: the unique map $\rho : G \longrightarrow \text{Aut } V$, where $\dim V=0$.

Less trivially, the group S_n of all permutations of the n elements $A_1, A_2, \ldots, A_n$ has a canonical n-dimensional representation. Let V be the vector space of dimension n with basis elements labeled $A_1, \ldots, A_n$. S_n acts on V by permuting the basis elements: given $\sigma \in S_n$ and $\sum \alpha_i A_i \in V$, $\sigma(\sum \alpha_i A_i) = \sum \alpha_i A_{\sigma(i)}$.

A permutation representation of a group G, of degree n, is by definition a homomorphism $G \longrightarrow S_n$. Given such, we can compose this map with the map $S_n \longrightarrow Gl(n, C)$ given above to get a linear representation $G \longrightarrow Gl(n, C)$. In particular, the group G acts on itself by left multiplication, giving a linear representation of G of degree $n = |G|$, called the regular representation of G.

Generalizing this process, we can start with any linear representation of a group G, and any group homomorphism $H \longrightarrow G$. Then composition of maps gives a linear representation of H. In particular, if H is a subgroup of G, each representation $\rho : G \longrightarrow \mathrm{Aut}\, V$ gives by restriction a representation of H, $\mathrm{Res}_G^H \rho : H \longrightarrow \mathrm{Aut}\, V$.

Conversely, if H is a subgroup of G, G acts on the set of left cosets of H in G, $\{gH \mid g \in G\}$ by left multiplication, giving a permutation representation of G, so by the above, a linear representation of G. Using a terminology which will later be generalized, we say that this representation of G is <u>induced</u> by the trivial one-dimensional representation of H.

A representation $\rho : G \longrightarrow \mathrm{Aut} V$ is <u>faithful</u> if is one-one. If not, let H be the kernel of ρ. Then ρ gives a faithful representation $G/H \longrightarrow \mathrm{Aut}\, V$. In this case, ρ will be said to be <u>associated</u> with the normal subgroup H. In particular, if ρ is induced by a subgroup H of G, the subgroup $\mathrm{Ker}(\rho)$ is contained in H and equals H if and only if H is normal.

Thus already for the group S_3, we have the following representations (where we just indicate $\rho(\sigma)$ and $\rho(\tau)$ since the other matrices are a consequence of these):

1) The trivial representation of degree n:

$$\rho(\sigma) = \rho(\tau) = \begin{pmatrix} 1 & & 0 \\ & 1 & \\ 0 & \cdots & 1 \end{pmatrix} \quad \text{the } n \times n \text{ unit matrix}$$

2) The canonical representation: Let $A_1=\begin{pmatrix}1\\0\\0\end{pmatrix}, A_2=\begin{pmatrix}0\\1\\0\end{pmatrix}, A_3=\begin{pmatrix}0\\0\\1\end{pmatrix}$, then

$$\rho(\sigma)=\begin{pmatrix}0&1&0\\1&0&0\\0&0&1\end{pmatrix}\qquad \rho(\tau)=\begin{pmatrix}0&0&1\\1&0&0\\0&1&0\end{pmatrix}$$

3) The regular representation: Let the elements of S_3 be taken in the order $1, \tau, \tau^2, \sigma, \sigma\tau, \sigma\tau^2$ then

$$\rho(\sigma) = \begin{pmatrix}0&0&0&1&0&0\\0&0&0&0&1&0\\0&0&0&0&0&1\\1&0&0&0&0&0\\0&1&0&0&0&0\\0&0&1&0&0&0\end{pmatrix}\qquad \rho(\tau) = \begin{pmatrix}0&0&1&0&0&0\\1&0&0&0&0&0\\0&1&0&0&0&0\\0&0&0&0&1&0\\0&0&0&0&0&1\\0&0&0&1&0&0\end{pmatrix}$$

4) The representation induced by the subgroup $H = \{1,\sigma\}$ of S_3. The cosets of H in S_3 are $H = \sigma H = \{1,\sigma\}$, $\tau H = \sigma\tau^2 H = \{\tau, \sigma\tau^2\}$, and $\tau^2 H = \sigma\tau H = \{\tau^2, \sigma\tau\}$.

$$\sigma(H) = H \qquad \tau(H) = \tau H$$

$$\sigma(\tau H) = (\sigma\tau)H \qquad \tau(\tau H) = (\tau^2)H$$

$$\sigma(\tau^2 H) = (\sigma\tau^2)H \qquad \tau(\tau^2 H) = H$$

Let V be the vector space with basis

$$H = \begin{pmatrix}1\\0\\0\end{pmatrix}\qquad \tau H = \begin{pmatrix}0\\1\\0\end{pmatrix}\qquad \tau^2 H = \begin{pmatrix}0\\0\\1\end{pmatrix}.$$

Then

$$\sigma = \begin{pmatrix}1&0&0\\0&0&1\\0&1&0\end{pmatrix} \text{ and } \tau = \begin{pmatrix}0&0&1\\1&0&0\\0&1&0\end{pmatrix}$$

This is isomorphic to the canonical representation but this is not so easy to see. Later, the theory of characters will make it trivial to check such an isomorphism.

5) The representation induced by the subgroup $K = \{1, \tau, \tau^2\}$ of S_3. This representation is two-dimensional. Letting $K = \begin{pmatrix} 1 \\ 0 \end{pmatrix}$ and $K = \begin{pmatrix} 0 \\ 1 \end{pmatrix}$ be a basis, we have

$$\sigma = \begin{pmatrix} 0 & 1 \\ 1 & 0 \end{pmatrix} \qquad \tau = \begin{pmatrix} 1 & 0 \\ 0 & 1 \end{pmatrix}$$

6) Another representation, not given by the above construction, is the <u>alternating</u> representation . It is of degree one, and assigns $\sigma = (-1)$ and $\tau = (1)$, assigning (± 1) depending on whether the permutation is even or odd.

Let G be a given group, and consider the class of all representations of G. There are a number of operations which can be performed on this class, and our object is to construct out of this class a ring R(G) with all of these operations built into the arithmetic of R(G).

Let V and W be two representations of G. The <u>sum</u> $V \oplus W$ of the two representations is constructed by taking the vector space sum $V \oplus W$ and letting G act by, for $g \in G$, $(v,w) \in V \oplus W$, $g(v,w) = (gv, gw)$. In terms of matrices, this is quite simple. Given bases $v_1, \ldots, v_n$

of V and $w_1, \ldots, w_n$ of W, the set $v_1, \ldots, v_n, w_1, \ldots, w_m$ is a basis of $V \oplus W$. Given $g \in G$, if A is the $n \times n$ matrix of G in Aut V with respect to the basis $v_1, \ldots, v_n$, and B the $m \times m$ matrix in Aut W with respect to $w_1, \ldots, w_m$, the matrix of g in Aut $V \oplus W$ with respect to the combined basis is $\begin{pmatrix} A & 0 \\ 0 & B \end{pmatrix}$ where 0 denotes the $n \times m$ and $m \times n$ zero matrices.

The <u>product</u> $V \cdot W$ of the two representations V and W is the vector space $V \otimes W$ on which $g \in G$ acts by $g(v \otimes w) = gv \otimes gw$. If $v_1, \ldots, v_n$ is a basis of V and $w_1, \ldots, w_m$ a basis of W, then the set of all symbols $v_i \otimes w_j$ is a basis of $V \otimes W$. The associated operation on matrices is called the Kroneckor product.

The ith <u>exterior power</u> $\lambda^i V$ is obtained by taking the ith exterior power of the vector space V and letting $g \in G$ act by $g(v_1 \wedge \ldots \wedge v_i) = gv_1 \wedge \ldots \wedge gv_i$. Note $\lambda^1 V = V$. We take $\lambda^0 V$ to be the trivial one-dimensional representation. If $n = \dim V$, $\lambda^n V$ is then the one-dimensional representation assigning to each $g \in G$, the determinant of the matrix assigned to g in the representation V. If $n > \dim V$, $\lambda^n V$ is the zero representation.

Given G-modules V,W we have already described how Hom(V,W) is to be considered as a G-module. In particular, if W is the trivial one-dimensional G-module, Hom(V,W) is the <u>dual</u> $\overline{V}$ of the representation V. G thus acts on $\overline{V}$ by, for $\xi \in \overline{V}$, $v \in V$, $g \in G$, $(g\xi)(v) = \xi(g^{-1}v)$. In terms of matrices, we take a basis of V, and the corresponding dual basis of $\overline{V}$. Then the matrix assigned to g acting on $\overline{V}$ is the inverse transpose of that assigned to g in V.

Let $\rho : G \to \operatorname{Aut} V$ be a linear representation. Pick a basis $v_1, \ldots, v_n$ of V. Then for each $g \in G$, $\rho(g)$ is an $n \times n$ matrix with complex number entries. Let $\rho^*(g)$ be the $n \times n$ matrix obtained from $\rho(g)$ by taking the conjugates of the entries. $\rho^*(g) \in \operatorname{Aut} V$, and thus $\rho^* : G \to \operatorname{Aut} V$ is another representation, denoted V^*, and called the <u>conjugate</u> representation of V. (Of course one must check that the transition $V \rightsquigarrow V^*$ does not depend, up to isomorphism of V^*, on the choice of basis.)

<u>Proposition</u>: For any representation V, $\overline{V}$ is isomorphic to V^*.

<u>Proof</u>: By the previous proposition, we can assume V has a G-invariant inner product. Pick an orthonormal basis of V. Then G acts on V via unitary matrices. But a unitary matrix is one whose inverse transpose is identical to its conjugate. ∎

Let 0 be the zero representation of G and 1 the trivial one-dimensional representation.

<u>Proposition:</u> Let U, V, and W be G-modules. Then there are isomorphisms:

i) $(U \oplus V) \oplus W \cong U \oplus (V \oplus W)$

ii) $V \oplus U \cong U \oplus V$

iii) $0 \oplus V \cong V \oplus 0 \cong V$

iv) $(U \cdot V) \cdot W \cong U \cdot (V \cdot W)$

v) $U \cdot V \cong V \cdot U$

vi) $U \cdot 1 \cong 1 \cdot U \cong U$

vii) $U \cdot (V \oplus W) \cong U \cdot V \oplus U \cdot W$

viii) $\overline{U \oplus V} \cong \overline{U} \oplus \overline{V}$

ix) $\overline{U \cdot V} \cong \overline{U} \cdot \overline{V}$

x) $\overline{\overline{U}} \cong U$

xi) $\mathrm{Hom}(U, V) \cong \overline{U} \cdot V$

xii) $\mathrm{Hom}(V, W) \cong \mathrm{Hom}(\overline{W}, \overline{V}) \cong \overline{\mathrm{Hom}(W, V)}$

xiii) $\lambda^n(V \oplus W) \cong \sum_{i=0}^{n} \lambda^i(V) \cdot \lambda^{n-i}(W)$

xiv) $\mathrm{Hom}(V, W_1 \oplus W_2) \cong \mathrm{Hom}(V, W_1) \oplus \mathrm{Hom}(V, W_2)$

xv) $\mathrm{Hom}(V_1 \oplus V_2, W) \cong \mathrm{Hom}(V_1, W) \oplus \mathrm{Hom}(V_2, W)$

Proof: All of these are well known to be true if both sides of each equivalence are considered as vector spaces. The only problem is to check that the defined action of G on both sides is the same. In each case, this is clear. ∎

We now construct the representation ring R(G) of the group G. Take the set of all finite formal sums $\sum_i n_i [V_i]$ where V_i are representations of G and n_i are integers. This set is a group under the operation $\sum n_i[V_i] + \sum m_i[V_i] = \sum(n_i+m_i)[V_i]$. We now require that

i) If V and W are isomorphic representations, $[V] = [W]$.

ii) For all V and W, $[V] + [W] = [V+W]$.

The resulting set is R(G).

Thus if an element $\sum n_i[V_i]$ has all its coefficients n_i positive, we can let $W = \bigoplus_i (V_i + \ldots + V_i)$ (where there are n_i summands of V_i). W is then a representation for which $W = \sum n_i[V_i]$. Such an element of R(G) is called an "actual representation". The general element $\sum n_i[V_i]$ can be written as a difference of two actual representations: $\sum n_i[V_i] = [W_1] - [W_2]$, where W_1 comes from those n_i which are positive, W_2 from those which are negative. A general element is often referred to as a "virtual representation" of G. Often we will be sloppy and say "the virtual representation W_1-W_2" for "the element $[W_1] - [W_2]$ of R(G)".

Parts i),ii),iii) of the above proposition imply that R(G) is an abelian group under +. Defining $[W]\cdot[U] = [W\cdot U]$, parts iv),v),vi),vii) show that R(G) is a commutative ring. Defining $\overline{[V]} = [\overline{V}] = [V^*]$, parts viii),ix),x) show that $\overline{}$ is an involution on R(G). Part xi) shows that the Hom operation does not introduce anything new. xii), a simple consequence of ix),x),xi) will be used later. Finally, xiii), together with the conventions on λ^0 and λ^1 imply that R(G) is a pre-λ- ring. Note that for any actual representation V, the dimension of V is equal to the degree of $\lambda_t(V)$.

R(G) is in fact a λ-ring, but this will be more easily seen later with character theory.

The equivalence of $\overline{V}$ and V* allows us to construct the following <u>inner product</u> on R(G):

$$(V,W) = \dim_{\mathbb{C}}\mathrm{Hom}_G(V,W)$$

The form is bilinear by xiv),xv) and nondegenerate, since for any V, $\mathrm{Hom}_G(V,V)$ includes at least the one-dimensional space of scalar multiples of the identity map. The subtle point is the symmetry of (-,-). A simple computation shows $\mathrm{Hom}_G(V^*,W^*) = (\mathrm{Hom}(V^*,W^*))^G = (\mathrm{Hom}(\overline{V},\overline{W}))^G = (\overline{\mathrm{Hom}(V,W)})^G = (\mathrm{Hom}(W,V))^G = \mathrm{Hom}_G(W,V)$. Hence it is sufficient to show that $\dim_{\mathbb{C}}\mathrm{Hom}_G(V,W) = \dim_{\mathbb{C}}\mathrm{Hom}_G(V^*,W^*)$. Pick bases of V and W. Consider the map $\mathrm{Hom}_G(V,W) \longrightarrow \mathrm{Hom}_G(V^*,W^*)$ defined by: given $f: V \to W$, write f as a matrix. Let f* be the matrix of conjugate entries, considered as a map $V^* \to W^*$. Then f is G-invariant if and only if f* is, setting up a one-one correspondence between these spaces. The correspondence is additive, and for any <u>real</u> number r, $r(f^*) = (rf)^*$. Hence the spaces are isomorphic as real vector spaces, so have the same dimension as complex vector spaces. Thus (-,-) is indeed an inner product.

As the simplest example of the above, let G be the trivial group. The resulting R(G) is then just the ring of integers, with $\overline{}$ the identity map, $\lambda^n(m) = \binom{m}{n}$, and $(n,m) = nm$, under the isomorphism $[V] \rightsquigarrow \dim V$.

Let G be a subgroup of H. We have already remarked in this case that every representation of H gives by restriction a representation of G. It is not hard to see that this gives a homomorphism of λ-rings with involution and augmentation $\mathrm{Res}: R(H) \longrightarrow R(G)$. In particular, if G is the trivial subgroup, $R(G) \cong \mathbb{Z}$, so this is the augmentation $R(H) \longrightarrow \mathbb{Z}$, $[V] \rightsquigarrow$ degree V.

Conversely, if G is a subgroup of H and V a representation of G, we can define the <u>induced representation</u> Ind(V) of H as follows: G acts on H as a set of permutations by $g(h) = hg$. Let $\mathbb{C}H$ be the vector space with basis the elements of H. This action of G makes $\mathbb{C}H$ a <u>right</u> G-module (i.e., $\rho: G \longrightarrow \mathrm{Aut}\,\mathbb{C}H$ is an anti-homomorphism: $\rho(gh) = \rho(h)\rho(g)$.) Ind(V) is then defined as the vector space $\mathbb{C}H \otimes V$, modulo the subspace generated by all elements of the form $hg \otimes v - h \otimes gv$. H acts on Ind(V) by, for $h_1 \in H$, $h_1(h \otimes v) = (h_1 h) \otimes v$. (Since $h_1(hg \otimes v) = hh_1 g \otimes v = h_1 h \otimes gv = h_1(h \otimes gv)$, this makes sense.) If several groups are involved, we sometimes write Ind_G^H for $\mathrm{Ind}: R(G) \longrightarrow R(H)$.

It is easy to see that $\mathrm{Ind}(V \oplus W) = \mathrm{Ind}(V) \oplus \mathrm{Ind}(W)$ so that Ind gives an additive map $R(G) \longrightarrow R(H)$. Slightly less trivially, $\mathrm{Ind}(\overline{V}) = \overline{\mathrm{Ind}(V)}$. To see this, note that as an H-module, $\mathbb{C}H = \mathbb{C}H^*$, since H acts by matrices with real entries. Hence $\mathrm{Ind}(V^*) = (\mathrm{Ind}(V))^*$, whence the conclusion.

Ind increases the augmentation: if $n = |H| / |G|$, then $\mathrm{degree}(\mathrm{Ind}_G^H(V)) = n$ degree V.

These operations Ind and Res are related by Frobenius Reciprocity:

<u>Theorem:</u> Let $G \subset H$. Let V be a representation of G, W of H.

1) $\dim_{\mathbb{C}} \mathrm{Hom}_G(V, \mathrm{Res}\, W) = \dim_{\mathbb{C}} \mathrm{Hom}_H(\mathrm{Ind}\, V, W)$

2) $\mathrm{Ind}(V \otimes \mathrm{Res}\, W) = (\mathrm{Ind}\, V) \otimes W$

In other words, 1) Res and Ind are adjoint operators on the inner product spaces R(G) and R(H), and 2) $\mathrm{Ind}(R(G)) \subset R(H)$ is an ideal.

<u>Proof:</u> There is a natural imbedding $V \longrightarrow \mathbb{C}H \otimes_{\mathbb{C}G} V$, $v \rightsquigarrow 1 \otimes v$. Consider the diagram of vector spaces

$$\begin{array}{ccc} V & \overset{f}{\dashrightarrow} & \mathrm{Res}\, W \\ \text{inclusion} \downarrow & & \downarrow \text{isomorphism of vector spaces} \\ \mathrm{Ind}\, V = \mathbb{C}H \otimes_{\mathbb{C}G} V & \underset{g}{\dashrightarrow} & W \end{array}$$

Given a G-map $f: V \longrightarrow \mathrm{Res}\, W$, there is a unique H-map g, making this diagram commute: take $g(h \otimes v) = hf(v)$. Given an H-map g, f is uniquely determined as the restriction of g to the subspace V of Ind V.

The isomorphism $\mathrm{Ind}(V\otimes \mathrm{Res}\, W) \to \mathrm{Ind}(V)\otimes W$ is defined by $h\otimes(v\otimes w) \rightsquigarrow (h\otimes v)\otimes hw$. The inverse map is $(h\otimes v)\otimes u \rightsquigarrow h\otimes(v\otimes h^{-1}u)$. Each is an H-module homomorphism. ∎

2. Irreducible Representations and Schur's Lemma

Let V be a representation of a group G. V is called <u>reducible</u> if there is a subspace $W \subset V$, with $gw \in W$ for all $w \in W, g \in G$ and with $W \neq \{0\}, V$. Otherwise V is <u>irreducible</u>. (The zero representation is by definition reducible.) V is called <u>decomposible</u> if V is isomorphic to the sum of two representations of G: $V \cong V_1 \oplus V_2$. Otherwise, V is <u>indecomposible</u>.

<u>Theorem</u> (Maschke): If V is reducible, V is decomposible.

<u>Proof</u>: In other words, if W is a nontrivial G-submodule of V, there is another G-submodule W' of V with $W \oplus W' \cong V$. To see this, choose a G-invariant inner product on V (as on page 62). Given W, let W' be its orthogonal complement. ∎

Let Irrep $G = \{V_1, V_2, \ldots\}$ be the set of (isomorphism classes) of irreducible representations of G. We take V_1 to be the trivial one-dimensional representation, which is irreducible for any group G.

Any representation V of G can be written as a finite sum of irreducibles: $V = \bigoplus_{\text{IrrepG}} n_i V_i$. To accomplish this, proceed as follows: If V is irreducible, $V = V_i$, for some $V_i \in$ Irrep G. If not, let $V_i \subsetneq V$ be a nontrivial G-submodule of V of minimal dimension. Then V_i is irreducible, and by the theorem $V \cong V_1 \oplus V_2$ where $V_2 \subset V$ is a G-module of smaller dimension than V. Proceed by induction.

<u>Schur's</u> <u>Lemma</u>: Let $V_i, V_j \in$ Irrep G. Then $\dim_{\mathbb{C}} \mathrm{Hom}_G(V_i, V_j)$ is one or zero according as to whether i=j or not. In particular, if $f: V_i \longrightarrow V_i$ is a G-module homomorphism, then there is a complex number λ such that $f(v) = \lambda v$ for all $v \in V_i$.

<u>Proof:</u> Suppose $f: V_i \longrightarrow V_j$. Then the kernel of f is a G-submodule of V_i so is either 0 or V_i. Similarly, the image of f is 0 or V_j. Hence if f is not the zero map, it must be both one-one and onto, hence an isomorphism. If $V_i = V_j = V$, $f: V \longrightarrow V$, let $v \in V$ be an eigenvector of f, with eigenvalue λ. (Such always exists since V is a complex vector space.) Consider the map $f - \lambda I: V \longrightarrow V$. Since its kernel is non-zero, (it contains v) and V is irreducible, $f - \lambda I$ is the zero map. Hence $f = \lambda I$. ∎

<u>Theorem</u>: The irreducible representations of G form an orthonormal basis of the representation ring R(G) with respect to the inner product $(-;-)$.

<u>Proof:</u> Schur's lemma gives the orthonormality immediately, and the comment after Maschke's Theorem shows that the irreducibles span R(G). ∎

Corollary: As an abelian group under + with inner product, R(G) is isomorphic to $\mathbb{Z}^k$, $k = |\text{Irrep } G|$, and $((n_1, n_2, \ldots), (m_1, m_2, \ldots)) = n_1 m_1 + n_2 m_2 + \ldots$. ∎

Corollary: Let $V \in R(G)$ be any element. Then V is $\pm$ an irreducible representation if and only if $(V,V)=1$. ∎

Corollary: Let $\mathcal{R}$ be the regular representation of G. Then $\mathcal{R} = \bigoplus_{\text{IrrepG}} n_i V_i$, where each $V_i \in \text{Irrep}(G)$ appears, and $n_i = \dim V_i$. In particular, the number of irreducible representations of G is finite. Also $|G| = \sum_{\text{IrrepG}} (\dim V_i)^2$.

Proof: Let 1_0 be the trivial one-dimensional representation of the unit subgroup 1 of G. Then $\mathcal{R} = \text{Ind}(1_0)$ so by Frobenius Reciprocity, for each irreducible representation V_i of G, $n_i = (\mathcal{R}, V_i) = (\text{Ind}(1_0), V_i) = (1_0, \text{Res } V_i) = \dim \text{Hom}_{\{1\}}(1_0, V_i) = \dim_{\mathbb{C}} V_i$. ∎

A representation V of G is called isotypical of type V_i, if it is of the form $V = nV_i$, for some irreducible representation V_i.

Proposition: Let $V_i \in \text{Irrep } G$. Given any G-module V, there is a unique maximal G-submodule W of V which is isotypical of type V_i of V. It is called the V_i-isotypical component of V. V is the sum of its isotypical components.

Proof: Write $V \cong \bigoplus_{\text{Irrep} G} n_i V_i$ as a sum of irreducibles. This gives an isomorphism $V \cong (n_i V_i) \oplus (\bigoplus_{j \neq i} n_j V_j)$. Under this isomorphism $n_i V_i$ corresponds to an isotypical G-submodule W of V, and $\bigoplus_{j \neq i} n_j V_j$ to a complementary submodule W' of V. We claim that any isotypical submodule of V of type V_i must be contained in W. If not, there would be a submodule of V, isomorphic to V_i, which intersects W' nontrivially. Since V_i is irreducible, this makes V_i a submodule of W'. Hence V_i is isomorphic to a submodule of $\bigoplus_{j \neq i} n_j V_j$, so $\bigoplus_{j \neq i} n_j V_j = V_i \oplus (\text{complementary part})$ so $0 = (\bigoplus_{j \neq i} n_j V_j, V_i) = (V_i \oplus \text{complement}), V_i) = (V_i, V_i) + (\text{complement}, V_i) = 1 + (\text{a non-negative integer})$. This contradiction gives the result. ■

In fact, one can be more explicit about the decomposition of V into isotypical components. For each $V_i \in \text{Irrep } G$, let $V_i \otimes \text{Hom}_G(V_i, V) \longrightarrow V$ be the map $v \otimes f \rightsquigarrow f(v)$. Then $\bigoplus_{\text{Irrep } G} (V_i \otimes \text{Hom}_G(V_i, V)) \longrightarrow V$ is an isomorphism and is the decomposition of G. Note that $\text{Hom}_G(V_i, V)$ is a trivial G-module. If we map $\mathbb{Z} \longrightarrow R(G)$ by $n \rightsquigarrow nV_i$, this makes $\text{Hom}_G(V_i, V) = n_i$.

<u>Proposition</u>: Let G be a subgroup of H, $V_{i_o} \in$ Irrep G and $W_{j_o} \in$ Irrep H. Then in writing $\mathrm{Ind}(V_{i_o}) = \bigoplus_{\mathrm{Irrep}H} m_j W_j$, and $\mathrm{Res}(W_{j_o}) = \bigoplus_{\mathrm{Irrep}G} n_i V_i$, the coefficients m_{j_o} and n_{i_o} are equal.

<u>Proof</u>: This is just a restatement of Frobenius Reciprocity. ∎

3. Characters

Let G be a finite group. A central function on G is a map of sets $f:G \longrightarrow \mathbb{C}$, such that $f(ab)=f(ba)$ for all a,b G. (Equivalently, $f(a)=f(b)$ whenever a and b are conjugate in G.) Let CF(G) be the set of all central functions on G. We define the following operations on CF(G):

Addition: $(f_1+f_2)(a) = f_1(a) + f_2(a)$

Zero: $0(a) = 0$

Multiplication: $(f_1 \cdot f_2)(a) = f_1(a)f_2(a)$

One: $1(a) = 1$

Dual: $\overline{f}(a) = f(a^{-1})$

Conjugate: $f^*(a) = (f(a))^*$

Adams Operations $(\Psi^n f)(a) = f(a^n)$

Inner Product $(f_1, f_2) = \frac{1}{|G|} \sum_{g \in G} f_1(g)^* f_2(g)$

Augmentation degree(f) = f(e)

As per page 54, CF(G) is a λ-ring under sum, product, and the λ-operations defined via the Adams operations.

CF(G) is also an algebra over $\mathbb{C}$. We can construct a natural basis as follows: For each conjugacy class K of G, let K denote also the function on G, $K(a)=1$ or 0, according to whether $a \in K$ or not. It is customary to order the conjugacy classes $K_1, K_2, \ldots, K_n$

taking K_1 to be the (singleton) class containing the unit element of G. Clearly, each K_i is a central function, and every central function f is of the form $f = \sum_i c_i K_i$, for unique $c_i \in \mathbb{C}$. Thus the dimension of CF(G) as $\mathbb{C}$-vector space is the number of conjugacy classes of G.

Lemma: For any pair of conjugacy classes K_i, K_j

$$(K_i, K_j) = \begin{cases} 0 & K_i \neq K_j \\ |K_i|/|G| & K_i = K_j \end{cases}$$

Proof: A simple computation. ∎

Since the K_i form an orthogonal set with respect to the bilinear form (-,-), it is an inner product.

Proposition: For all $f \in CF(G)$

1) $\psi^{|G|} f(a) = \text{degree}(f)$ for all $a \in G$
2) $\psi^{|G|+n}(f) = \psi^n(f)$ for all integers $n > 0$
3) $\psi^{|G|-1}(f) = \bar{f}$

Proof: Again a simple computation, using the fact that $a^{|G|} = e$ for all $a \in G$. ∎

Before giving the main theorem in this section, we recall some facts about the trace of an automorphism. Let $\alpha : V \longrightarrow V$ be a linear transformation of a vector space V into itself. Pick a basis of V and let $A = (A_{ij})$ be the matrix of α. The trace of α, $Tr(\alpha)$, is defined as $\sum_i A_{ii}$, the sum of the diagonal entries. It has the following properties:

Lemma: i) $Tr(\alpha)$ is well-defined; it does not depend on the choice of basis of V

ii) If $\alpha : V \longrightarrow W$ and $\beta : W \longrightarrow V$, then $Tr(\alpha\beta) = Tr(\beta\alpha)$.

iii) If $\alpha : V \longrightarrow V$ is the identity map, $Tr(\alpha) = \dim V$. If α is the zero map, $Tr(\alpha) = 0$.

iv) Given $\alpha : V \longrightarrow V$, $\beta : W \longrightarrow W$ so that $\alpha \oplus \beta : V \oplus W \longrightarrow V \oplus W$, $Tr(\alpha \oplus \beta) = Tr(\alpha) + Tr(\beta)$.

v) Given $\alpha : V \longrightarrow V$, $\beta : W \longrightarrow W$ so that $\alpha \otimes \beta : V \otimes W \longrightarrow V \otimes W$, $Tr(\alpha \otimes \beta) = Tr(\alpha) Tr(\beta)$.

vi) Given $\alpha : V \longrightarrow V$, $Tr(\alpha) = Tr({}^t\alpha)$ (${}^t\alpha$ = transpose α)

vii) Given $\alpha : V \longrightarrow V$, $Tr(\alpha) = Tr(\alpha^*)$ (α^* = conjugate α)

viii) Let $\alpha : V \longrightarrow V$, inducing $\lambda^i_\alpha : \lambda^i V \longrightarrow \lambda^i V$, $i \geq 1$. Let T be a formal symbol (a "variable"). $\mathrm{Det}(I + \alpha T) = \sum_i Tr(\lambda^i \alpha) T^i$. (The left-hand side here is evaluated by picking a basis for V, taking the matrix for in that basis, multiplying each entry by T, adding the identity matrix, and taking the determinant of the result)

ix) If $\alpha : V \to V$ is idempotent (i.e., $\alpha \cdot \alpha = \alpha$) then $\mathrm{Tr}(\alpha) = \dim(\mathrm{Image}\,\alpha)$. $\blacksquare$

Let $\rho : G \to \mathrm{Aut}\, V$ be a representation of a group G. We define a map of sets $\chi_{(V,\rho)} : G \to \mathbb{C}$ (usually denoted just χ_V) by $\chi_{(V,\rho)}(g) = \mathrm{Tr}(\rho(g))$, $g \in G$. By part ii) of the lemma above, χ_V is a central function so χ_V is an element of CF(G), the <u>character</u> of the representation V. Also by ii), χ_V depends only on the isomorphism type of (V, ρ). By iii), $\chi_{V \oplus W} = \chi_V + \chi_W$. Hence the assignment of characters to representations gives a map $R(G) \to CF(G)$.

<u>Theorem</u>: The map $R(G) \to CF(G)$ is a homomorphism of λ-rings preserving the involution, conjugation, inner product, and augmentation. It is a one-one map, identifying R(G) with its image, which we call the <u>character ring</u> of G. The images of the irreducible representations, called the <u>irreducible characters</u>, form an orthonormal basis of CF(G).

<u>Proof</u>: The fact that the map preserves sum, product, 1, 0, the involution, the conjugate, and the augmentation follow immediately from the lemma above.

To show that the map preserves the λ-ring structure, it is sufficient to show that it preserves the Adams operations. Thus, given a G-module V, if we take the element $\psi^k(V) \in R(G)$, computed via the universal polynomial $\psi^k(V) = Q_k(\lambda^1 V, \lambda^2 V, \ldots, \lambda^k V)$, and then compute $\chi_{\psi^k(V)}$, the result must be the same as computing $\psi^k(\chi_V)$. Hence we must show, for every $g \in G$, $\chi_{\psi^k(V)}(g) = \psi^k(\chi_V)(g)$. Let $g \in G$ and assume a basis for V is picked so that the matrix for g is diagonal (which is possible by).

$$g = \begin{pmatrix} g_1 & & & 0 \\ & g_2 & & \\ & & \ddots & \\ 0 & & & g_n \end{pmatrix}$$

Then, using the lemma above:

$$\begin{aligned} 1 + \operatorname{Tr}(g)T + \operatorname{Tr}(\lambda^2 g)T^2 + \ldots &= \operatorname{Det}(I + gT) \\ &= \operatorname{Det}\begin{vmatrix} 1+g_1T & & & 0 \\ & 1+g_2T & & \\ & & \ddots & \\ 0 & & & 1+g_nT \end{vmatrix} \\ &= \prod (1+g_iT) \end{aligned}$$

Thus $\operatorname{Tr}(\lambda^i g)$ is the ith elementary symmetric function of the g_i's.

We now compute

$$\begin{aligned} \chi_{\psi^k(V)}(g) &= Q_k(\operatorname{Tr}\lambda^1 g, \operatorname{Tr}\lambda^2 g, \ldots, \operatorname{Tr}\lambda^k g) \\ &= g_1^k + g_2^k + \ldots + g_n^k \qquad \text{(by definition of } Q_k) \\ &= \operatorname{Tr}(g^k) \\ &= \chi_V(g^k) \\ &= \psi^k \chi_V(g) \end{aligned}$$

Hence the map $R(G) \to CF(G)$ preserves the λ-ring structure.

Next we show that the map preserves the inner product. This will have two other consequences: the map must be one-one, and the set of irreducible characters is an orthonormal set in CF(G). First we need a fact: given any representation V of G,

$$\frac{1}{|G|}\sum_{g \in G} \chi_V(g) = \dim V^G .$$

Proof: Let $J: V \longrightarrow V$ be the linear map defined by

$$J = \frac{1}{|G|}\sum_{g \in G} g .$$

Then the composition gJ is equal to J for all $g \in G$. It is a simple consequence that J is idempotent and every $g \in G$ acts trivially on the image of J. Hence Image(J) $\subset V^G$. But also, J is the identity map on V^G, so $V^G \subset$ Image(J). Hence V^G = Image(J) so by the lemma, $\dim V_G = \mathrm{Tr}(J) = \frac{1}{|G|}\sum_{g \in G} \mathrm{Tr}(g)$. QED.

Now take any two representations V and W of G, with characters χ_V, χ_W.

$$\begin{aligned}
(\chi_V, \chi_W) &= \frac{1}{|G|}\sum_{g \in G} \chi_V(g)^* \chi_W(g) \\
&= \frac{1}{|G|}\sum_{g \in G} \chi_{V^*}(g) \chi_W(g) \\
&= \frac{1}{|G|}\sum_{g \in G} \chi_{\bar{V}}(g) \chi_W(g) \\
&= \frac{1}{|G|}\sum_{g \in G} \chi_{\bar{V} \otimes W}(g) \\
&= \frac{1}{|G|}\sum_{g \in G} \chi_{\mathrm{Hom}(V,W)}(g) \\
&= \dim_{\mathbb{C}} \mathrm{Hom}_G(V,W) \qquad \text{(using the fact above)} \\
&= (V,W)
\end{aligned}$$

Finally, we show that the set of irreducible characters span $CF'G)$. Let $f \in CF'G)$ be any central function. Let $\rho: G \longrightarrow \text{Aut } V$ be an irreducible representation of degree n. Define $\rho_f = \sum_{g \in G} f(g)\rho(g)$, a map from V to V. Then $\rho_f = \lambda I$, a constant times the identity map, where $\lambda = (|G|/n)(f^*, \chi_V)$. Proof: Let $g \in G$. Then

$$\begin{aligned}
\rho(g)^{-1}\rho_f\rho(g) &= \sum_{g_1 \in G} f(g_1)\rho(g)^{-1}\rho(g_1)\rho(g) \\
&= \sum_{g_1 \in G} f(g_1)\rho(g^{-1}g_1g) \\
&= \sum_{g_2 \in G} f(gg_2{}^{-1}g)\rho(g_2) \quad \left(\text{where } g_2 = g^{-1}g_1g\right) \\
&= \sum_{g_2 \in G} f(g_2)\rho(g_2) \quad \left(f \text{ is central so } f(gg_2{}^{-1}g) = f(g_2)\right) \\
&= \rho_f \quad .
\end{aligned}$$

Hence, by Schur's lemma, ρ_f, is a constant multiple of the identity: $\rho_f = \lambda I$. Calculating the trace of both sides, we get $n\lambda = \text{Tr}(\lambda I)$ $= \sum_{g \in G} f(g)\text{Tr}(\rho(g)) = \sum_{g \in G} f(g)\chi_V(g) = (f^*, \chi_V)$. QED.

Now let f be any central function on G. Suppose f is orthogonal to all irreducible characters, i.e., $(f, \chi_V) = 0$ for all irreducible V. Hence for every irreducible representation $\rho: G \longrightarrow \text{Aut } V$, ρ_{f^*} is zero, by the above. Since every representation is a sum of irreducibles, ρ_{f^*} is zero for all $\rho: G \longrightarrow \text{Aut } V$. Now apply this to the regular representation. We calculate the transform of the basis vector e by

ρ_{f*}: $0 = \rho_{f*}(e) = \sum_{g \in G} f(g)^* \rho_g(e) = \sum_{g \in G} f(g)^* g$. Thus $f(g)^* = 0$ for all $g \in G$, since the set of all $g \in G$ form a basis of the regular representation. Hence $f = 0$.

Given any central function f, now, $\left[f - \sum_{\mathrm{IrrepG}} (f, \chi_{V_i})\chi_{V_i}\right]$ is orthogonal to all irreducible characters, so zero.

Hence $f = \sum_{\mathrm{IrrepG}} (f, \chi_{V_i})\chi_{V_i}$. ∎

Corollary: R(G) is a λ-ring.

Proof: We already knew R(G) to be a pre-λ-ring. In question now is whether a number of identities hold. But since R(G) is contained as a pre-λ-ring in CF(G) and the identities hold in CF(G) the conclusion follows. ∎

Since $\overline{V} = V^*$ in R(G), this gives a necessary (but hardly sufficient) condition for a central function to be a character: $\chi_V(g^{-1}) = \chi_{\overline{V}}(g) = \chi_{V^*}(g) = \chi_V(g)^*$. For another proof of this identity, take coordinates in V so that g is a diagonal matrix:

$$g = \begin{pmatrix} g_1 & & 0 \\ & \ddots & \\ 0 & & g_n \end{pmatrix}$$

Since $g^{|G|} = e$, each g_i must be a $|G|^{\text{th}}$ root of unity. Hence, for each i, $g_i^* = g_i^{-1}$, so $\mathrm{Tr}(g^*) = \mathrm{Tr}(g^{-1})$.

Corollary: The number of irreducible representations of G is equal to the number of conjugacy classes of G. ■

Corollary: G is abelian if and only if every irreducible representation of G has dimension one. Equivalently, G is abelian if and only if for every representation V of G, a basis for V can be found so that all the matrices assigned to G by that basis are diagonal.

Proof: We already know that the number of irreducible representations of G, n, is the number of classes of G, and the sum of the squares of the degrees is G: $\sum_{i=1}^{n} d_i^{2} = |G|$, integer d_i. Hence each $d_i=1$ if and only if $n = |G|$. The second assertion of the corollary follows immediately from the first. ■

We now have two natural bases for R(G) - the irreducible characters χ_V, and the class functions K_i. We index the irreducible characters by upper indices: χ^i, letting χ^1 be the character of the trivial one-dimensional representation which is irreducible for any group G. The symbol χ^i_j refers to that complex number obtained when the character χ^i is applied to an element of K_j.

Since both the sets χ^i and K_j are orthogonal for the inner product, we can easily express either in terms of the other:

$$\chi^i = \sum_j \frac{(\chi^i, K_j)}{(K_j, K_j)} K_j$$

$$K_j = \sum_i \frac{(K_j, \chi^i)}{(\chi^i, \chi^i)} \chi^i$$

Computing, we find $(\chi^i, K_j) = (K_j, \chi^i)^* = \frac{|K_j|}{|G|} \chi_j^{i\,*}$. Recall $(K_j, K_j) = \frac{|K_j|}{|G|}$.

Hence $\chi^i = \sum_j \chi_j^{i*} K_j$.

If we renormalize the K_j by defining $L_j = \frac{|G|}{|K_j|} K_j$, the L_j are also an orthonormal basis:

$$(L_i, L_j) = \begin{cases} 0 & i \neq j \\ \frac{|G|}{|K_i|} & i=j \end{cases}$$

and $(\chi^i, L_j) = \chi_j^i$. Thus $L_j = \sum_j \chi_j^i \chi^i$.

<u>Proposition</u>: $\sum_i \chi^i(g)^* \chi^i(h) = \frac{|G|}{|K|}$ if g and h are conjugate and both contained in the class K; zero, otherwise.

Proof: Say $g \in K_i$ and $h \in K_j$. Define L_i, L_j as above. Then the sum in question is $\sum_k \chi_i^{k*} \chi_j^k = (L_i, L_j)$. ∎

(Note that in general, K_j and L_j are only central functions, not characters.)

The matrix of complex numbers χ_j^i is called the <u>character table</u>

of the group G. It is traditionally written listing the classes across the top and irreducible characters at the left:

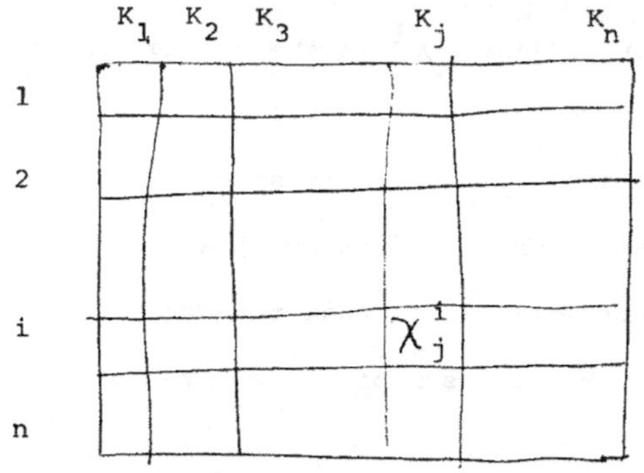

The construction of this table for a given group G is the basic calculation of the representation theory of the group. Since we will need to construct a number of examples later, it is worthwhile to review the main features of the table.

The table has two interpretations. As above, it represents in two ways linear transformations between orthogonal bases, and thus the rows and columns obey the <u>orthogonality relations</u> given above:

rows: $$\frac{1}{|G|}\sum_{g\in G}\chi^{i}(g)^{*}\chi^{j}(g) = \delta_{ij} = (\chi^{i},\chi^{j})$$

columns: $$\sum_{i\in \mathrm{IrrepG}}\chi^{i*}_{j}\chi^{i}_{k} = \delta_{jk}\frac{|G|}{|K_j|}$$

Also, there are as many rows as columns, the common number being the number of conjugacy classes in G.

There is also the interpretation of the table as the listing, for each irreducible representation of G, of the values taken by the corresponding character. Because of this, certain features of the table are obvious. Since χ^1 is the trivial one-dimensional character, the first row consists of all ones. Since K_1 is the trivial class, the first column is a list of the degrees of the irreducible representations. A fact which we will not prove here (but see Serre, Theorem II.12) is that the degrees divide $|G|$. We have already proved that the sum of the squares of the degrees of the irreps is G .

Another useful observation is that a one-dimensional representation of G (necessarily irreducible) is just a group homomorphism $\rho : G \longrightarrow \mathbb{C}$. Since $\mathbb{C}$ is abelian, ρ factors through G modulo its commutator subgroup $[G,G]$, inducing a representation ρ' of $G/[G,G]$.

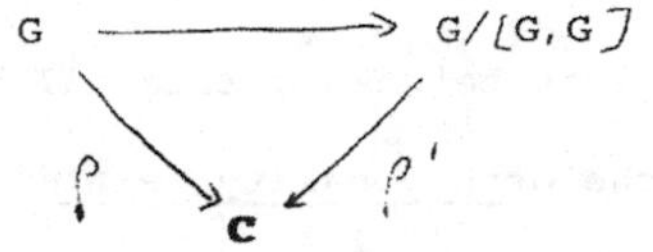

Conversely, any one-dimensional representation of $G/[G,G]$ gives an irreducible representation of G. Hence, in particular, the number of irreducible one-dimensional characters of G is the order of $G/\ G,G$. Indeed, the character table of $G/\ G,G$, or for that matter G/H for any normal subgroup H, appears as a part of the character table of G.

Since the representation ring of the group G is isomorphic to the character ring, the addition, multiplication, conjugacy, and λ-operations can all be carried out on the characters, and it is possible to be very explicit about the arithmetic of R(G).

As an example, we shall describe $R(S_3)$ completely. S_3 has three conjugacy classes: $K_1 = \{e\}$, $K_2 = \{\tau, \tau^2\}$, and $K_3 = \{\sigma, \sigma\tau, \sigma\tau^2\}$. Hence S_3 has three irreducible representations. Since $|S_3| = 6$, the sum of the squares of the degrees of these irreps must be 6. Hence the degrees must be 1,1 and 2. The commutator subgroup of S_3 is $\{1, \tau, \tau^2\}$ and $S_3/[S_3, S_3]$ is the cyclic group of 2 elements $\mathbb{Z}_2$. It is easy to write down the character table for $\mathbb{Z}_2$. Hence we already have the first two rows of the table. Orthogonality of columns gives the last row.

	K_1	K_2	K_3
χ^1	1	1	1
χ^2	1	1	-1
χ^3	2	-1	0

χ^2 is the character of the alternating representation of S_3. χ^3 is the character of the matrix representation given on page 60.

Every representation of S_3 is an integral combination of these three. For example, consulting the list on pp.65-7 , the trivial representation of degree n has character (n,n,n) so is $n\chi^1$. The regular representation has character (6,0,0) so is $\chi^1 + \chi^2 + 2\chi^3$, in agreement with the corollary . The representation induced by the subgroup $\{1, \sigma\}$ of S_3 is (3,0,1) so it is $\chi^1 + \chi^3$. The canonical representation has character (3,0,1) so also is $\chi^1 + \chi^3$, thus proving the isomorphism mentioned on page 7. Etc.

Let us now write out explicitly the arithmetic of $R(S_3)$. Additively, $R(S_3)$ is the free abelian group on three generators χ^1, χ^2, χ^3. Since the characters take all real values, the involution is trivial. $\text{Degree}(n_1\chi^1+n_2\chi^2+n_3\chi^3) = n_1+n_2+2n_3$. $(n_1\chi^1+n_2\chi^2+n_3\chi^3, m_1\chi^1+m_2\chi^2+m_3\chi^3) = n_1m_1+n_2m_2+n_3m_3$.

To compute the multiplication, we can just compute $\chi^i\chi^j$, for each pair (i,j). For example, the product of the two characters χ^2 and χ^3 is the character

$$(\chi^2\chi^3)(e) = \chi^2(e)\cdot\chi^3(e) = 1\cdot 2 = 2$$
$$(\chi^2\chi^3)(\tau) = \chi^2(\tau)\chi^3(\tau) = -1$$
$$(\chi^2\chi^3)(\sigma) = \chi^2(\sigma)\chi^3(\sigma) = 0$$

Hence $\chi^2\chi^3$ is the character $(2,-1,0)$ so equal to χ^3.

The full multiplication table for $\chi^i\chi^j$ is

	χ^1	χ^2	χ^3
χ^1	χ^1	χ^2	χ^3
χ^2	χ^2	χ^1	χ^3
χ^3	χ^3	χ^3	$\chi^1+\chi^2+\chi^3$

The general product of two elements of $R(G)$ follows by bilinearity.

χ^1 and χ^2 are one-dimensional, so the λ- operations are easy on them: $\lambda_t(\chi^1) = 1 + \chi^1 t$, $\lambda_t(\chi^2) = 1 + \chi^2 t$. χ^3 is two-dimensional so $\lambda_t(\chi^3) = 1 + \chi^3 t + at^2$ for some $a \in R(G)$. a can be computed in two ways. First we have matrices for χ^3, and a is just the character obtained from the second exterior power of these matrices; in this case this is the determinant since they are 2×2. Or, we can compute $\psi^2(\chi^3)$ by

$$\psi^2(\chi^3(e)) = \chi^3(e^2) = \chi^3(e) = 2$$

$$\psi^2(\chi^3(\tau)) = \chi^3(\tau^2) = \chi^3(\tau) = -1$$

$$\psi^2(\chi^3(\sigma)) = \chi^3(\sigma^2) = \chi^3(e) = 2$$

so $\psi^2(\chi^3)$ is the character (2,-1,2). Now, in any λ-ring, $\lambda^2(x) = (1/2)(\psi^1(x)\psi^1(x) - \psi^2(x))$. $\psi^1(\chi^3) = \chi^3 = (2,-1,0)$. It's square, using the formula for product of two central functions, is (4,1,0). Subtracting $\psi^2(\chi^3) = (2,-1,2)$ and dividing the result by two yields $\lambda^2(\chi^3) = (1,1,-1) = \chi^2$.

Exercise: Carry out the same calculation for the two non-abelian groups of order eight (the quaternion group and the dihedral group). Show the resulting character rings are isomorphic as rings, but not isomorphic as λ-rings. (Hint on the latter: calculate in each case how many solutions there are to the equation $\lambda^2(x) = 1$).

Let H be a subgroup of G and χ a character of a representation of H. We compute a formula for the induced character $\mathrm{Ind}_H^G \chi$. Let π be a conjugacy class of G, and L_π the associated central function (p. 90). Recall, for any character χ of G, the dot product $(\chi, L_\pi) = \chi(\pi)$. Thus

$$\begin{aligned} \mathrm{Ind}_H^G\chi(\pi) &= {}_G(\mathrm{Ind}_H^G\chi, L_\pi) \\ &= {}_H(\chi, \mathrm{Res}\, L_\pi) \qquad \text{(by Frobenius Reciprocity)} \\ &= \frac{1}{|H|} \sum_{\sigma\in H} \chi(\sigma)\, \mathrm{Res}\, L_\pi(\sigma) \end{aligned}$$

(here using the fact that L_π takes real values so $L_\pi(\sigma) = L_\pi(\sigma)^*$).

By definition

$$\mathrm{Res}L_\pi(\sigma) = L_\pi(\sigma) = \begin{cases} \frac{G}{[\pi]} & \sigma\in\pi \\ 0 & \sigma\notin\pi \end{cases}$$

so $$\mathrm{Ind}_H^G\chi(\pi) = \frac{|G|}{|H|}\frac{1}{[\pi]} \sum_{\sigma\in H\cap[\pi]} \chi(\sigma)$$

In particular, if χ is the unit character,

$$\mathrm{Ind}_H^G\, 1(\pi) = \frac{|G|}{|H|}\frac{|H\cap[\pi]|}{[\pi]}$$

(where, recall, the notation $[\pi]$ refers to both the conjugacy class π, and the number of elements of this class).

The next computation is the representation ring $R(G\times H)$ of the cartesian product of two groups. Suppose $\rho_1:G\longrightarrow \mathrm{Aut}\ V_1$ and $\rho_2:H\longrightarrow \mathrm{Aut}\ V_2$ are given representations. Let $\rho_1\times\rho_2:G\times H\longrightarrow \mathrm{Aut}(V_1\otimes V_2)$ be defined by

$$\rho_1\times\rho_2((g,h))(v_1\otimes v_2) = \rho_1(g)(v_1)\otimes\rho_2(h)(v_2) \quad .$$

This is a representation and its character $\chi_{\rho_1\times\rho_2}$ is given by

$$\chi_{\rho_1\times\rho_2}((g,h)) = \chi_{\rho_1}(g)\ \chi_{\rho_2}(h)$$

(since the trace of a tensor product is the product of the traces.)

<u>Proposition</u>: If ρ_1 and ρ_2 are irreducible, so is $\rho_1\times\rho_2$. Conversely, every irrep of $G\times H$ is given in this way.

<u>Proof</u>: We have given

$$\frac{1}{|G|}\sum_{g\in G}|\chi_{\rho_1}(g)|^2 = 1 \qquad\qquad \frac{1}{|H|}\sum_{h\in H}|\chi_{\rho_2}(h)|^2 = 1 \quad .$$

Hence, using the formula for $\chi_{\rho_1\times\rho_2}$ above,

$$\frac{1}{|G\times H|}\sum_{(g,h)\in G\times H}|\chi_{\rho_1\times\rho_2}(g,h)|^2 = 1$$

so, by the corollary on p.78, $\chi_{\rho_1\times\rho_2}$ is irreducible. Similarly, if ρ_1,ρ_1' are irreps of G, and ρ_2,ρ_2' of H, and either $\rho_1\neq\rho_1'$ or $\rho_2\neq\rho_2'$, then $\rho_1\times\rho_2\neq\rho_1'\times\rho_2''$.

To check that $\{\rho_1 \times \rho_2 \mid \rho_1 \in \text{IrrepG}, \rho_2 \in \text{IrrepH}\}$ is a complete set of irreps of $G \times H$, one only need compute

$$\sum_{\substack{\rho_1 \in \text{IrrepG} \\ \rho_2 \in \text{IrrepH}}} (\text{degree}(\rho_1 \times \rho_2))^2 = \sum_{\rho_1, \rho_2} (\deg \rho_1 \cdot \deg \rho_2)^2$$

$$= \sum_{\rho_1, \rho_2} (\deg \rho_1)^2 (\deg \rho_2)^2 = \sum_{\rho_1} (\deg \rho_1)^2 \sum_{\rho_2} (\deg \rho_2)^2$$

$$= |G| \, |H| = |G \times H| \qquad \blacksquare$$

There are other ways to compose two groups to get a third, and corresponding theorems on the associated representation rings. Given groups G, H and a homomorphism $G \xrightarrow{\varphi} \text{Aut } H$ (where Aut here refers to automorphisms of H as a group) the <u>semi-direct product</u> $H \underset{\varphi}{\times} G$ is the set of pairs $(h,g) \in H \times G$, with the multiplication rule $(h,g)(h',g') = (h\,\varphi(h'), gg')$. For the construction of the characters of a semidirect product in the special case that H is abelian, see Serre ([41] , p.II-18).

More complex is the <u>wreath product</u> of a permutation group G and a group H. Here there is given an action, $G \longrightarrow \text{Aut } S$, of G on a set S with n elements. This induces a group homomorphism $G \longrightarrow \text{Aut } (H \times \ldots \times H)$ (n factors) by $g(\ldots,h_s,\ldots) = (\ldots,h_{g^{-1}(s)},\ldots)$.

The associated semidirect product of $H^{\times n}$ with G is the wreath product, written G[H]. These products will arise later with the special case of $G = S_n$, the symmetric group on n letters.

To finish this section, we give without proof three standard results in representation theory, and some corollaries. While none of this is necessary for the sequal, no account of subject would be complete without at least their mention. (The proofs are not so difficult - just irrelevant here).

The first is Mackey's Theorem. Let G be a group and H and K be subgroups. Let $\rho: H \to \mathrm{Aut}\, W$ be a representation and $\rho' = \mathrm{Res}_G^K \mathrm{Ind}_H^G \rho$ be the representation obtained by inducing ρ up to G, and then restricting back to K. The problem is to compute ρ'.

Choose a set of double coset representatives of G mod (H,K) - i.e., a set $S \subset G$ such that G is the disjoint union of KsH, $s \in S$. (A standard notation is $S = K\backslash G/H$.) For $s \in S$, let $H_s = sHs^{-1} \cap K$, a subgroup of K. Put, for $x \in H_s$, $\rho_s(x) = \rho(s^{-1}xs)$, giving a representation $\rho_s : H_s \to \mathrm{Aut}\, W$. $H_s \subset K$ so $\mathrm{Ind}_{H_s}^K(\rho_s)$ is defined.

<u>Theorem</u>: The representation $\mathrm{Res}_G^K \mathrm{Ind}_H^G(\rho)$ is isomorphic to the sum of the representations $\mathrm{Ind}_{H_s}^K(\rho_s)$ for $s \in K\backslash G/H$.

(For the proof, and some corollaries, see Serre (41 ,p.II-10). For a categorical generalization to other functors than $G \rightsquigarrow R(G)$, see A.Dress (15)).

Corollary: Let H,K be subgroups of G. Let $\varphi_H = \mathrm{Ind}_H^G 1$, $\varphi_K = \mathrm{Ind}_K^G 1$. Then the dot product (φ_H, φ_K) is equal to the number of double cosets of G mod (K,H). ∎

The second theorem, already mentioned above, is that for any group G, and any irrep $\rho : G \to \mathrm{Aut}\, V$ of G, the degree of ρ divides the order of G. This is usually proved (e.g., in Serre (41 , p.II.4)) via the theorem that the values of the characters $\chi_\rho(g)$, $g \in G$, are "algebraic integers". A more natural proof would be as a corollary to the following conjecture:

Conjecture: Let G be a finite group and $\rho \in R(G)$ an irreducible representation. Let $\rho_G \in R(G)$ be the regular representation of G. Then there is an element $\xi \in R(G)$ with $\xi\rho = \rho_G$. (In particular $\deg \xi \deg \rho = \deg \rho_G = |G|$, hence the statement above.)

This conjecture is true in many examples, but unproved in general.

The third theorem is Brauer's Theorem on induced characters.

Theorem: Let G be a finite group. Every character on G is a linear combination, with integer coefficients, of characters induced up from characters of degree 1 on various subgroups of G. (For the proof, see Serre ([41] , p.II-29)).

The reason we mention this is to point out a fact which is, at least, not evident in most of the literature on representation theory: the character table of a finite group is effectively ("recursively") computable, given the multiplication table of the group. In particular, it follows that the character ring R(G) is effectively describeable. Not only can one say R(G) is, as an abelian group, free on a certain number of generaters, one can give explicitly a set of generaters as functions $\chi_i : G \to \mathbb{C}$, and describe explicitly the products $\chi_i\chi_j = \sum_k r_{ij}^k \chi_k$, and λ-operations $\lambda^q(\chi_i) = \sum_q s_{iq}\chi_q$, giving the integers r_{ij}^k, s_{iq}.

We will prove the effectivity of the character table by describing a (admittedly inefficient) algorithm for its computation.

First observe that, for any finite group G, the characters of degree 1 are effectively computable. Indeed each such is a group homomorphism $\chi : G \to \mathbb{C}$. If G has order n, so $g^n = e$, all $g \in G$, then $(\chi(g))^n = \chi(g^n) = \chi(e) = 1$ so the image of χ is contained in the (finite) group μ_n of nth roots of unity in $\mathbb{C}$. Hence one-dimensional characters are group homomorphisms $G \to \mu_n$, and since both groups are finite, a computer can easily list all homomorphisms between them.

Now let G be any group. Given a subgroup H of G, compute all the one-dimensional characters χ of H, and then compute the associated induced characters $\mathrm{Ind}_H^G\chi$ of G, again a mechanical procedure. Doing this for every $H \subset G$, we get a finite set $\{\chi_q\}$ of representations of G. Set the computer to listing all integer combinations of these χ_q. E.g., first list all such combinations with integer coefficients of absolute value ≤ 1, then those with coefficients of absolute value ≤ 2, etc.

The important facts are now:

i) By Brauer's theorem, each irrep of G will eventually appear on this list.

ii) Given any linear combination of the χ_q, it is a mechanical computation to see if χ is an irrep of G: test if $(\chi,\chi)=1$ and $\chi(e)>0$.

iii) Given two irreps, χ_1 and χ_2, it is a mechanical matter to check if they are isomorphic: test if $(\chi_1,\chi_2)=0$ or 1.

iv) It is known in advance how many irreps G has - namely, compute the number of conjugacy classes of G.

Thus the computer generates a list of representations, checks each to see if it is irreducible, and stops when the right number of irreps is found.

It would be nice if Brauer's Theorem could be strengthened to assert that every irrep of a group G is induced from a one-dimensional representation of some subgroup of G. Groups having this property are called <u>monomial</u>. Alas, not all groups are monomial. An example of a non-monomial group is the semidirect product of the quaternion group $H = \{\pm 1, \pm i, \pm j, \pm k\}$ with the cyclic group $\mathbb{Z}/3\mathbb{Z}$ acting on H by cyclicly permuting i,j,k. This product has an irrep of degree 2, but no subgroup at all of index 2.

4. Permutation Representations and the Burnside Ring

Let G be a group (finite as always) and S a finite set. An action of G on S is a homomorphism of G to the group of permutations of S. For $g \in G$ and $s \in S$, we write g(s) for the element of S obtained by applying the permutation associated to g, to the element s. For $g_1, g_2 \in G$, we have $g_1(g_2(s)) = (g_1 g_2)(s)$.

S is called a G-set. If two G-sets, S and T, are given, a G-map is a map of sets $f: S \longrightarrow T$ such that $f(g(s)) = g(f(s))$, for all $g \in G$. f is an isomorphism and S and T are isomorphic, if f is a G-map which is an isomprphism as map of sets - i.e., one-one and onto.

If S has n elements, the map $G \longrightarrow \text{Aut } S$ is a map $G \longrightarrow S_n$, and so the concept of G-set is the same as that of permutation representation of degree n of G, discussed above. Note that we do not require the map $G \longrightarrow S_n$ to be one-one!

The object of this section is to mimic the linear representation theory of groups and the construction of R(G) with the construction of the Burnside ring B(G) for the permutation representation theory of groups.

An orbit of the action of G on S is any subset of S of the form $\{gs \mid g \in G\}$ for some $s \in S$. An orbit is clearly determined by any element in it, and S is the disjoint union of orbits. The action is <u>transitive</u> , or S is a <u>simple</u> G-set, if all of S is one orbit. Equivalently, given any $s_1, s_2 \in S$, there is a $g \in G$ with $g(s_1)=s_2$.

For one example of a G-set, consider the action of G on itself by inner multiplication (=conjugation). To avoid confusion here, lets write φ_g for the permutation associated to g. The action is then defined by, for $g,h \in G$, $\varphi_g(h) = g^{-1}hg$. Note that φ_g is a group homomorphism. Such automorphisms of G are called <u>inner</u> automorphisms.

G also acts on the set of subgroups of G by conjugation. Given $g \in G, H \subset G$, take $\varphi_g(H) = \{ g^{-1}hg \mid g \in G,\ h \in H\}$. Some terminology - given $H \subset G$, $\{ g \in G \mid \varphi_g(H) \subset H\}$ is called the <u>normalizer</u> of H, and $\{g \in G \mid \varphi_g(h)=h, \text{ for all } h \in H\}$ is the <u>centralizer</u> of H.

Earlier, a class of examples of permutation representations of G was described. Given a subgroup H of G, G acts on the set of left cosets $G/H = \{gH \mid g \in G\}$ of H in G by left multiplication: $g_1(gH) = (g_1 g)H$. This action is transitive and we have the converse:

Proposition: Every transitive permutation representation of G is of the form G/H for some subgroup H of G.

Proof: Suppose G acts transitively on S. Pick $s \in S$ and let $H = \{g \mid g(s)=s\}$. Consider the G-set G/H and the map $G/H = \{gH \mid g \in G\} \xrightarrow{\varphi} S$ given by $\varphi(gH) = gs$. It is easily checked that φ is well-defined and a G-map and an isomorphism. ∎

Corollary: If G acts transitively in S, then the number of elements of S divides the number of elements of G. ∎

Note what this says about the action of G on itself by inner multiplication. In that case, the orbit of an element is its conjugacy class, and so we have that the number of elements conjugate to a given element divides the order of the group.

Let S,T be G-sets. Then under the obvious action, the disjoint union of S and T is a G-set, called the sum of S and T and written S+T.

G also acts on the cartesian product, $S \times T$, by $g(s,t)=(g(s),g(t))$, giving the product $S \times T$ of S and T.

For each integer $n \geq 1$, G acts on the n-fold symmetric power of T. This set is constructed by taking all n-tuples of elements of T, T^n, and identifying for any permutation $\sigma \in S_n$,

$(t_1,t_2,\ldots,t_n) \sim (t_{\sigma(1)},\ldots,t_{\sigma(n)})$. G acts by g(the equivalence class containing $(t_1,\ldots,t_n)$ = the equivalence class of $(g(t_1),\ldots,g(t_n))$. We write $\mathrm{Symm}^n T$ for this G-set.

As in the case of linear representations we construct a ring out of all permutation representations. The <u>Burnside</u> <u>ring</u>, B(G), of a group G, consists of all finite formal sums, $\Sigma_i [S_i]$, of G-sets S_i, modulo the relation: $[S_1]+[S_2]$, the sum in B(G), is equal to $[S_1+S_2]$. The product, as defined above, makes B(G) a ring, and the symmetric power operation, interpreted as operations h_n, give a λ-ring structure to B(G).

(The fact that B(G) is a λ-ring and not just a pre-λ-ring - i.e., the truth of all the indentities - follows from the fact that these identities hold amongst sums, products, and symmetric powers of ordinary sets, and hold "naturally". Another proof - more analogous to our proof of λ-hood for R(G) - would be via the embedding of B(G) into a "ring of characters". But the necessary relevant lemma for "characters" and B(G) is not yet known. See below (p.113).

Given any G-set S, S is obviously the sum, as G-sets, of orbits. Hence for permutation representations, the analog of Maschke's Theorem is easy (making the analogy: irreducible≡simple).

Proposition: Every G-set S can be written as a finite sum of transitive G-sets, $S = S_1 + S_2 + \ldots + S_n$. Furthermore, the set $\{S_i\}$ of transitive G-sets appearing is unique (up to order). ∎

Hence, as abelian group, B(G) is free of rank k, where k is the number of non-isomorphic transitive G-sets. To compute k, we need the analog of Schur's Lemma.

Lemma: Let U and V be subgroups of G. Then $\mathrm{Hom}_G(G/U, G/V)$, the set of G-maps from G/U to G/V, is nonempty if and only if some conjugate $g^{-1}Ug$ of U is contained in V.

Proof: Let $\varphi: G/U \longrightarrow G/V$. Let $e \in G$ be the identity element. Pick $g \in G$ so that $gV = \varphi(eU)$. For any $h \in U$, $hU = U = eU$. Since φ is well-defined, $gV = \varphi(eU) = \varphi(hU) = h\varphi(U) = hgV$. Hence $g^{-1}hgV = V$, so $g^{-1}hg \in V$. This holds for all $h \in U$, so $g^{-1}Ug \subset V$. ∎

This is not quite as nice as the Schur Lemma - there can be different irreducible G-sets with a nontrivial map between. But one can characterize isomorphism of simple G-sets.

Corollary: Suppose G/U and G/V are isomorphic as G-sets. Then U and V are conjugate subgroups of G.

Proof: Since there is a map $G/U \longrightarrow G/V$, $g^{-1}Ug \subset V$ for some $g \in G$. The inverse isomorphism gives $h^{-1}Vh \subset U$, some $h \in G$. Hence $g^{-1}h^{-1}Vhg \subset g^{-1}Ug \subset V$. But $(hg)^{-1}Vhg$ has the same (finite!) number of elements as V. Hence $g^{-1}h^{-1}Vhg = V$, so $g^{-1}Ug = V$. ∎

Corollary: Rank(B(G)) = k = the number of conjugacy classes of subgroups of G. ∎

As in the first section of this chapter, to each permutation representation of G, we can associate a linear representation of G. This gives a homomorphism of λ-rings $B(G) \longrightarrow R(G)$.

It should be noted that this map is not usually one-one. E.g., for $G = S_3$, there are four conjugacy classes of subgroups but only three conjugacy classes of elements. Hence, as groups, $B(S_3) = \mathbb{Z}^4$ but $R(S_3) = \mathbb{Z}^3$, and so the map here can't be one-one.

Consider the composite map $B(G) \longrightarrow R(G) \longrightarrow CF(G)$, assigning to each permutation representation of G, the character of the associated linear representation.

Proposition: Let S be a G-set and χ the character of the associated linear representation. Then for $g \in G$, $\chi(g)$ is the number of points of S left fixed by g. In particular, χ is an integer-valued central function on G, taking only nonnegative values.

Proof: Immediate from the construction of the character of a representation. ∎

Matrices such as occur in such representations (choosing the obvious basis of the vector space involved) are permutation matrices each row and column contain only one non-zero entry, that entry being "1".

Hence the map $B(G) \longrightarrow R(G)$ has image at most the integer-valued characters of $R(G)$, and so isn't onto in general. In fact it isn't even onto the set of integer-valued characters (see, for example, the group discussed on p. 103).

So far, all of this is a straightforward generalization to permutation representations of the theory for linear representations. The first new ingredient is the analog of character theory. Since the composite $B(G) \longrightarrow R(G) \longrightarrow CF(G)$ is not one-one, usual characters do not suffice to distinguish unequal elements of $B(G)$. So the notion must be extended.

<u>Definition</u>: A <u>super</u> <u>central</u> <u>function</u> on G is a map of sets

$$\varphi: (\text{conjugacy classes of subgroups of } G) \longrightarrow \mathbb{C} \quad .$$

The set of super central functions is a ring (by $(\varphi_1+\varphi_2)([U]) = \varphi_1([U])+\varphi_2([U])$, $(\varphi_1\varphi_2)([U])=\varphi_1([U])\varphi_2([U])$) denoted $SCF(G)$.

Given a G-set S, the <u>super</u> <u>character</u> of this permutation representation is the function φ_S, given by, for each subgroup H of G, $\varphi_S(H)$ = the number of elements of $\{s \in S \mid h(s)=s, \text{ all } h \in H\}$. In shorthand notation, $\varphi_S(H) = |H^S|$.

Note that $\varphi_{S_1+S_2}(U) = \varphi_{S_1}(U)+\varphi_{S_2}(U)$ and $\varphi_{S_1S_2}(U)=\varphi_{S_1}(U)\varphi_{S_2}(U)$. Hence the map $B(G)\longrightarrow CSF(G)$ is a ring homomorphism.

One interpretation of $\varphi_S(U)$ is the number of elements in $Hom_G(G/U,S)$. The notation generalizes to apply to any G-set T, $\varphi_S(T) = |Hom_G(T,S)|$. Of course it is easy to compute $\varphi_S(T)$, given the values of $\varphi_S(U)$, all $U\subset G$.

<u>Theorem</u>: Two G-sets S and T are isomorphic if and only if the associated super characters φ_S and φ_T are equal.

<u>Proof</u>: One direction is of course trivial. To prove the other, it is convenient to set up first a partial ordering on the set of simple G-sets, using any of the following equivalent definitions:

$G/U < G/V$ if and only if $Hom_G(G/U,G/V) \neq \emptyset$

if and only if $g^{-1}Ug \subset V$ some $g\in G$

if and only if $\varphi_{G/V}(U) \neq 0$.

We write $U < V$ if $G/U<G/V$.

Now suppose $S = \sum n_U(G/U)$ and $T = \sum m_U(G/U)$ are given G-sets, where the sum ranges over a set of conjugacy classes of subgroups of G with integral coefficients n_U, m_U.

Suppose for all $V \subset G$, $\varphi_S(V) = \varphi_T(V)$. Then $0 = \varphi_S(V) - \varphi_T(V) = \sum(n_U-m_U)\varphi_{G/U}(V)$, for all $V\subset G$. Suppose for some $U \subset G$, $n_U \neq m_U$. Let V_0 be a maximal such (in the sense of the above ordering). Then $n_U - m_U = 0$ for $U > V_0$. But $\varphi_{G/U}(V_0) = 0$ unless $U \leq V_0$, so the sum becomes $0 = (n_{V_0} - m_{V_0})\varphi_{G/V_0}(V_0)$. Since $\varphi_{G/V_0}(V_0) \neq 0$, $n_{V_0} = m_{V_0}$, a contradiction. Hence $n_U = m_U$ for all $U \subset G$, so $S = T$. ∎

The categorical reader should notice here that the Yoneda Lemma exactly states that knowledge of the functor $T \rightsquigarrow \mathrm{Hom}(T,S)$ determines S up to isomorphism. This theorem says that we don't have to know the whole functor, just its effect on objects!

<u>Corollary</u>: The map $B(G) \longrightarrow SCF(G)$ is one-one. ∎

Given a supercentral function φ, and an element $g \in G$, φ can be applied to the cyclic subgroup generated by G.

<u>Lemma</u>: This assignment gives a ring homomorphism $SCF(G) \longrightarrow CF(G)$.

<u>Theorem</u>: There is a commutative diagram of maps of rings

$$\begin{array}{ccc} B(G) & \longrightarrow & R(G) \\ \downarrow & & \downarrow \\ SCF(G) & \longrightarrow & CF(G) \end{array}$$

∎

What's missing here is the λ-ring aspect of this construction. The obvious λ-structure to put on SCF(G) is to take as Adams operations, $\Psi^n(\varphi(H)) = \varphi(H^n)$, where H^n is the subgroup of G generated by n^{th} powers of elements of H. Then SCF(G) is a λ-ring (p. 54) and the bottom line is a λ-homomorphism. What remains to be proved is the conjecture : with this λ-structure on SCF(G), the map $B(G) \longrightarrow SCF(G)$ is a λ-homomorphism.

Dot products, also, have not been mentioned . The map $B(G) \longrightarrow R(G)$ isn't one-one, so that one cannot find any dot product on B(G) preserved by the map. But taking the transitive G-sets as an orthonormal basis of B(G) would define a dot product on B(G), having an easy extension to SCF(G). We leave it to the reader to figure out what the map $B(G) \longrightarrow R(G)$ has to do with the dot products involved.

Just as in linear representation theory, one can construct super character tables. Here are a couple.

Let G be a cyclic group of prime power order p. Conjugacy classes of subgroups of G are G itself, and {e}.

classes of subgroups

		{e}	G
simple G-sets	G/G	1	1
	G/{e}	|G|	0

Let $G = S_3$, with subgroups S_3, $\{e\}$, and the cyclic subgroups C_2, C_3 of orders 2 and 3.

classes of subgroups

		$\{e\}$	C_2	C_3	S_3
	S_3/S_3	1	1	1	1
simple G-sets	S_3/C_3	2	0	2	0
	S_3/C_2	3	1	0	0
	$S_3/\{e\}$	6	0	0	0

5. The Group Algebra Approach

Let G be a finite group. The group algebra of G, $\mathbb{C}[G]$, is the set of sums $\sum_{g\in G} c_g g$, $c_g \in \mathbb{C}$. $\mathbb{C}[G]$ is a vector space over $\mathbb{C}$, of dimension $|G|$, under $\sum c_g g + \sum d_g g = \sum (c_g + d_g) g$ and $c\sum c_g g = \sum (cc_g) g$, and an algebra, under

$(\sum c_g g)(\sum d_h h) = \sum_{gh=k} (c_g d_h) k$ (the multiplication induced by the group operation.

Given a representation $\rho : G \longrightarrow \text{Aut } V$, there is induced a $\mathbb{C}$-algebra homomorphism, also denoted ρ, $\rho : \mathbb{C}[G] \longrightarrow \text{End } V$, by, $\rho(\sum c_g g) = \sum c_g \rho(g)$.

Theorem: Let $\rho_i : G \longrightarrow \text{Aut } V_i$, $i \in I$, be a complete set of irreducible representations of G. Then the sum $\rho = \oplus \rho_i : \mathbb{C}[G] \longrightarrow \prod_{i\in I} \text{End } V_i$ is an isomorphism.

Proof: Suppose there is an element $\sum_{g\in G} c_g g \in \mathbb{C}[G]$ with $\rho(\sum c_g g) = 0$. Then $\rho_i(\sum c_g g) = 0$ for each i, so that $\sum c_g \rho_i(g)$ is the zero map on V_i, for each i. Since I runs through a set of all irreps of G, $\sum c_g g$ must induce the zero map on every G-module. In particular, $\mathbb{C}[G]$ is itself a G-module (by left multiplication) so $(\sum c_g g)\gamma = 0$ for all $\gamma \in \mathbb{C}[G]$. Take $\gamma = 1e$. Then $\sum c_g g = 0$, and the linear independence of the g's implies that $c_g = 0$, all g. Hence ρ is one-one. But since $\mathbb{C}[G]$ and $\prod \mathrm{End}\, V_i$ have the same dimension (Corollary, p. 78) ρ is also onto, so an isomorphism. ∎

This theorem reveals the connection between our approach and the traditional approach to representation theory. There, one starts with the algebra $\mathbb{C}[G]$ and observes that it is <u>semi-simple</u>: for any $\mathbb{C}[G]$-module M, and submodule N_1, there is a submodule N_2 with $M = N_1 \oplus N_2$. For $\mathbb{C}[G]$ this is, of course, the Maschke Theorem. Then one invokes Wedderburn's Theorem, a special case of which is that any semisimple $\mathbb{C}$-algebra is a sum of matrix algebras. Since the representation theory of a sum of matrix algebras is transparent (over $\mathbb{C}$!) this produces the main facts about representations of G.

From this point of view, a representation (group homomorphism) of G, $G \longrightarrow \mathrm{Aut}\, V$, is the same as a representation ($\mathbb{C}$-algebra homomorphism) of $\mathbb{C}[G]$, $\mathbb{C}[G] \longrightarrow \mathrm{End}\, V$. The reason that we have avoided this approach (aside from Occam's razor) is that in forming R(G), we want to take tensor products and exterior powers or representations.

Consider a vector space V and linear transformations $f,g,h: V \longrightarrow V$ inducing $f^n, g^n, h^n: \wedge^n V \longrightarrow \wedge^n V$ (by $f^n(v_1 \wedge \cdots \wedge v_n) = f(v_1) \wedge \cdots \wedge f(v_n)$ and similarly for g^n, h^n). Suppose $f = gh$, the composite of maps. Then $f^n = g^n h^n$. I.e., Λ^n is a functor. But Λ^n is not "additive". Even if $f=g+h$, i.e., $f(v)=g(v)+h(v)$, all $v \in V$, it can still happen that $f^n \neq g^n + h^n$ (as almost any example shows). Thus a map of groups, $G \longrightarrow \mathrm{Aut}\, V$, induces a map of groups $G \longrightarrow \mathrm{Aut}\, \Lambda^n V$, but a map of $\mathbb{C}$-algebras, $\mathcal{Q} \longrightarrow \mathrm{End}\, V$ doesn't induce a map of $\mathbb{C}$-algebras, $\mathcal{Q} \longrightarrow \mathrm{End}\, \Lambda^n V$.

Of course the exterior powers are not the only problem. Given representations $\mathcal{Q} \longrightarrow \mathrm{End}\, V$, $\mathcal{Q} \longrightarrow \mathrm{End}\, W$ of a $\mathbb{C}$-algebra there is no natural $\mathbb{C}$-algebra map $\mathcal{Q} \longrightarrow \mathrm{End}(V \otimes W)$.

Another indication of the problem is that a pair of nonisomorphic groups G_1, G_2 may yield isomorphic algebras $\mathbb{C}[G_1] = \mathbb{C}[G_2]$. For example, if G is abelian, $\mathbb{C}[G]$ is, using the theorem above and the fact that irreps of abelian groups are all one-dimensional, isomorphic to $\mathbb{C}^G$. Hence two abelian groups of the same order have isomorphic group algebras though their representation rings may differ. (E.g., this is true for the Klein 4-group and the cyclic group of order 4.) Or take the dihedral group D_4, and the group of quaternion units H. Again by the theorem above, and the known character tables of these groups, their group algebras are each isomorphic to $\mathbb{C} \oplus \mathbb{C} \oplus \mathbb{C} \oplus \mathbb{C} \oplus \mathrm{End}(\mathbb{C}^2)$. In fact as rings $R(D_4)=R(H)$. But as λ-rings, they differ (as pointed out on p. 95).

Let G be a group. By the theorem above, $\mathbb{C}[G]$ is isomorphic to a sum of matrix rings, $\prod_{i\in \mathrm{IrrepG}} \mathrm{End}\, V_i$, we can ask for those elements $e_i \in \mathbb{C}[G]$, $i \in \mathrm{IrrepG}$, satisfying $e_i = 1$ on V_i, 0 on and V_j, $j \neq i$. These elements are uniquely characterized by the properties

i) $e_i \gamma = \gamma \cdot e_i$ all $\gamma \in \mathbb{C}[G]$, all i (centrality)

ii) $e_i \cdot e_j = 0 \quad i \neq j$ (orthogonality)

$e_i \cdot e_i = e_i$ (idempotence)

iii) $\sum_i e_i = 1$ (completeness, positivity)

<u>Theorem</u>: Let χ^i be the ith irrep of G. Then

$$e_i = \frac{\deg \chi^i}{|G|} \sum_{g\in G} \chi^i(g)^* g \quad .$$

<u>Proof</u>: Let $Y_i \in \mathbb{C}[G]$ denote the element given by the above formula. For each G-module, $\rho: G \longrightarrow \mathrm{Aut}\, V$, Y_i acts on V by, for $v \in V$,

$$Y_i(v) = \frac{\deg \chi^i}{|G|} \sum \chi^i(g)^* \rho_g(v) \quad .$$

The trace of this endomorphism can be calculated as

$$\mathrm{Tr}_V(Y_i) = \frac{\deg \chi^i}{|G|} \sum \chi^i(g)^* \, \mathrm{Tr}(\rho_g)$$

$$= \mathrm{degree}\, \chi^i \; (\chi^i, \rho) \quad .$$

Furthermore the map $Y_i : V \longrightarrow V$ commutes with ρ_h for each $h \in G$:

$$\rho_h Y_i(v) = \frac{\deg \chi^i}{|G|} \sum_{g \in G} \chi^i(g)^* \rho_h(\rho_g(v))$$

$$= \frac{\deg \chi^i}{|G|} \sum_{q \in G} \chi^i(h^{-1}q)^* \rho_q(v) \qquad \text{(here taking } q = hg) \quad .$$

and

$$Y_i \rho_h(v) = \frac{\deg \chi^i}{|G|} \sum_{g \in G} \chi^i(g)^* \rho_g(\rho_h(v))$$

$$= \frac{\deg \chi^i}{|G|} \sum_{q \in G} \chi^i(qh^{-1}) \rho_q(v) \qquad \text{(here taking } q = gh) \quad .$$

Since χ^i is a central function, these two sums are equal.

Hence, by Schur's Lemma, if V is irreducible, $Y_i : V \longrightarrow V$ is a scalar multiple of the identity: $Y_i = rI$. Applying the calculation of the trace, $\mathrm{Tr}(Y_i) = r \cdot \deg V = \deg \chi^i \cdot (\chi^i, V)$. If $V = \chi^i$, r=1 and Y_i is the identity. If $V \neq \chi^i$, r=0. Hence Y_i is the projection operator which applied to any representation V of G, projects onto $Y_i(V) \subset V$, the χ^i-isotypical component. Since this is what the e_i's obviously do, the two must be identical. ■

The operators Y_i, for $i \in \mathrm{Irrep}\ G$, are called <u>Young Symmetrizing operators</u> (after the Rev. Alfred Young who first constructed them in the algebras $\mathbf{C}[S_n]$ - see [39], [47])

One application of these operators is in the simplest case of the symmetrizer, Y_0, of the unit representation of a symmetric group S_n. Let V be any vector space. Then the n-fold symmetric power of V has two equivalent definitions:

i) $\mathrm{Symm}^n V$ = the subspace of $V\otimes\ldots\otimes V$ (n factors) left fixed by the action of S_n (permuting factors).

ii) $\mathrm{Symm}^n V$ = the quotient of $V^{\otimes n}$ by W, where W is the subspace $W = \{ (x-\sigma x) \mid x\in V^{\otimes n}, \sigma\in S_n\}$.

The natural identification of these comes from the fact that, as subspace of $V^{\otimes n}$, $\mathrm{Symm}^n V$ is the image of Y_0. By the First Isomorphism Theorem $\mathrm{Image}(Y_0) = \mathrm{Domain}(Y_0)/\mathrm{Kernal}(Y_0)$ and we have $\mathrm{domain}(Y_0) = V^{\otimes n}$, $\mathrm{Ker}\ Y_0=W$. Similarly the nth exterior power, and any other type of power constructed out of an irrep of S_n, appears both as subspace and quotient of $V^{\otimes n}$.

We next prove a technical proposition which will be useful later. Given a Young symmetrizing operator Y, lets write Y_V for the endomorphism given by the action of $Y:V \longrightarrow V$, and Y_V' for the associated onto map $V \longrightarrow Y(V)$.

<u>Lemma</u>: Let Y be a Young symmetrizing operator of G. Given G-modules V,W the linear transformation $Y_{V\oplus W}: V\oplus W \longrightarrow V\oplus W$ is the sum, as linear transformation, of $Y_V: V \longrightarrow V$ and $Y_W: W \longrightarrow W$.

<u>Proof</u>: This is a simple corollary of the fact that every G-module is uniquely a sum of isotypical components. ∎

Lemma: Given a G-map $T:V \longrightarrow W$ of G-modules, there is a unique linear transformation T' so that the following commutes:

$$\begin{array}{ccc} V & \xrightarrow{\quad T \quad} & W \\ {\scriptstyle Y'_V}\big\downarrow & & \big\downarrow{\scriptstyle Y'_W} \\ Y(V) & \xrightarrow{\quad T' \quad} & Y(W) \end{array}$$

Proposition: Let $H \subset G$ and V be an H-module. Then there is a natural isomorphism

$$V^H \longrightarrow (\operatorname{Ind}_H^G V)^G \qquad .$$

Proof: Recall V^H denotes the elements of V left fixed by each $h \in H$, thus the image of the Young operator Y associated with the unit representation 1_H of H. Similarly for the superscript G. Consider the composite map

$$V^H \longrightarrow V \longrightarrow \operatorname{Ind}_H^G V \xrightarrow{\quad Y \quad} (\operatorname{Ind}_H^G V)^G$$

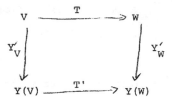

where the first map is inclusion of an isotypical subspace and the second is $V \longrightarrow \mathbb{C}[G] \underset{\mathbb{C}[H]}{\otimes} V$, $v \leadsto 1 \otimes v$. This composite map is an isomorphism! Indeed, by the first lemma, and the additivity of the formation of induced representations, it is additive in V, so we can restrict our attention to the case of irreducible V. If V is not the unit representation of H, $V^H = 0$ and also $(\operatorname{Ind} V)^G =$ $(\operatorname{Ind} V, 1_G)1_G = (V, \operatorname{Res}_G^H 1_G)1_G = (V, 1_H)1_G = 0\ 1_G = 0$ and any map from 0 to 0 is an isomorphism. In case $V = 1_H$, $V^H = V$, and (by Frobeni reciprocity) $(\mathrm{Ind}V)^G = (1_G + (\mathrm{other}))^G = 1_G$ and so the composite is a map $1_H \longrightarrow 1_G$. Hence it is sufficient to show that the map isn't zero. This follows easily by unravelling the definitions of induced representation and Young operator in this case. The final details are left to the reader as an exercise. ∎

Here is a final historical note. The original investigations of group representation theory were done, not with semisimple algebras, but with the socalled group determinant. Given a group G, one computes its group determinant, Θ, a polynomial function in variables X_g, $g \in G$. Given the "generic element $\sum_{g \in G} X_g g$ in $\mathbb{C}[G]$" consider the matrix assigned this element in the regular representation of G (in the obvious choice of basis of that space). Θ is the determinant of this matrix. The problem was to relate the structure of G to the problem of factoring Θ. Looking at this from our point of view, we see that the irreducible factors of Θ correspond to the irreducible representations of G, each factor occuring the number of times as its degree. It is interesting that an original impetus to the theory was Dedekind's conjecture (communicated to Frobenius in 1896) that the number of linear factors of Θ is equal to the index in G of its commutator subgroup (a simple fact for us - see p.92).

For a history of the early work in representation theory see Hawkins ([22]).

CHAPTER III : THE FUNDAMENTAL THEOREM

1. The Fundamental Theorem of the Representation Theory of the Symmetric Group

The theorem in question asserts that there is an isomorphism of λ-rings between the λ-ring Λ discussed in Chapter I, and a certain λ-ring constructed by combining all the representation rings of all symmetric groups.

Before we construct this object, it is necessary to recall a few facts about the symmetric group S_n of permutations of n letters, say the letters $1,2,3,\ldots,n$. Let σ be an element of S_n. σ is a <u>cycle of length q</u> if there is a subset $\{i_1,\ldots,i_q\} \subset \{1,\ldots,n\}$ such that $\sigma(i_1)=i_2, \sigma(i_2)=i_3,\ldots,\sigma(i_q)=i_1$ and $\sigma(m)=m$, for m not one of the i_j. Write $\sigma = (i_1,\ldots,i_q)$. Observe that every element $\sigma \in S_n$ can be written as a product of disjoint cycles: $\sigma=(i_1,\ldots,i_q)(j_1,\ldots,j_r)\ldots$ $\ldots(k_1,\ldots,k_s)$ – indeed under the action of σ, the set $\{1,\ldots,n\}$ breaks up into a number of orbits, each of which is a cycle.

The <u>cycle structure</u> of σ is the associated partition $\pi = (\lambda_1,\ldots,\lambda_n)$ where the λ_i are the sizes of the cycles in the decomposition. For example, the elements of S_3 have one of three different cycle structures:

$\pi = (1,1,1)$ corresponds to the element $e = (1)(2)(3)$

$\pi = (2,1)$ corresponds to the elements $\sigma = (12)(3)$, $\sigma\tau = (23)(1)$, and $\sigma\tau^2 = (13)(2)$

$\pi = (3)$ corresponds to the elements $\tau = (123)$ and $\tau^2 = (132)$

Proposition: Two elements of S_n are conjugate if and only if they have the same cycle structure.

Proof: Suppose σ and σ' have the same cycle structure. Write each as a product of disjoint cycles in the following pattern: one above the other, so that cycles of equal length correspond:

$$\sigma \qquad (i_1,\ldots,i_q)(j_1,\ldots,j_r)\ldots(k_1,\ldots,k_s)$$

$$\sigma' \qquad (i'_1,\ldots,i'_q)(j'_1,\ldots,j'_r)\ldots(k'_1,\ldots,k'_s)$$

Let $\tau \in S_n$ be the correspondence given by the vertical lines: $\tau(i_1)=i'_1,\ldots,\tau(i_q)=i'_q,\tau(j_1)=j'_1,\ldots$etc. Then $\tau^{-1}\sigma'\tau=\sigma$ so σ and σ' are conjugate. The converse proposition is clear. ∎

Corollary: The number of conjugacy classes of S_n, and hence also the number of irreducible representations of S_n, is equal to the number of partitions of n. ∎

Henceforth we adopt the notation $[\pi]$, for π a partition of n, for the conjugacy class of S_n of elements of cycle structure π.

The other calculation we will need is the number of elements of $[\pi]$ (which number will also be denoted by $[\pi]$.).

<u>Proposition</u>: Suppose $\pi = (1^{\alpha}2^{\beta}\ldots n^{\gamma})$ is a partition of n. Then

$$[\pi] = \frac{n!}{1^{\alpha}\alpha!\,2^{\beta}\beta!\ldots n^{\gamma}\gamma!}$$

<u>Proof</u>: Given the cycle structure π:

$$\underbrace{(-)(-)\ldots(-)}_{\alpha}\ \underbrace{(-,-)(-,-)\ldots(-,-)}_{\beta}\ldots\ldots\ldots\underbrace{(\overbrace{-,-,\ldots,-}^{n})}_{\gamma}$$

let us count the number of ways to fill in the blanks. There are n! ways to write in the numbers 1,2,...,n. However the first α one-cycles can be rearranged in $\alpha!$ ways giving the same element of S_n, so we must divide by $\alpha!$ to remove the duplication. Similarly the $\beta!,\ldots,\gamma!$ terms arise. Furthermore in each of the cycles of length q, any one of the q elements can be listed first - so we have to divide by q for each cycle of length q. Hence the factor $1^{\alpha}2^{\beta}\ldots n^{\gamma}$. ∎

We now construct the λ-ring R(S). For each integer n=0,1,2,... let $R(S_n)$ be the representation ring of S_n (where, recall, $R(S_0)=\mathbb{Z}$). For the while, forget the multiplication and λ-structure on each $R(S_n)$ and consider it as just an abelian group.

The outer product is a pairing $R(S_n)\times R(S_m) \longrightarrow R(S_{n+m})$ defined as follows. First observe that $S_n \times S_m$ can be considered as a subgroup of S_{n+m} by taking S_n to be permutations of 1,2,...,n, S_m as permutations of n+1,n+2,...,m, and S_{n+m} as permutations of 1,2,...,m. In dividing up the n+m symbols permuted by S_{n+m}

into one set of n and another set of m, there is some ambiguity - the division can be done in $\binom{n+m}{n}$ ways, but any two such ways of constructing an injection $S_n \times S_m \longrightarrow S_{n+m}$ give conjugate subgroups of S_{n+m}.

Given now elements $\alpha_n \in R(S_n)$, $\alpha_m \in R(S_m)$, consider the element $\alpha_n \times \alpha_m \in R(S_n \times S_m)$ (see definition. p.). The <u>outer product</u> $\alpha_n \alpha_m$ of α_n and α_m is the element of $R(S_{n+m})$ given by

$$\alpha_n \alpha_m = \mathrm{Ind}_{S_n \times S_m}^{S_{n+m}} (\alpha_n \times \alpha_m)$$

(Note that since any two of our ways of imbedding $S_n \times S_m$ into S_{n+m} are conjugate, the element $\alpha_n \alpha_m \in R(S_{n+m})$ is independent of the way chosen.) For n=0, the map $R(S_n) \times R(S_m) \longrightarrow R(S_{n+m})$ is just $(n,\alpha) \rightsquigarrow \alpha+\alpha+..+\alpha$ (n summands), and ditto for m=0.

In terms of actual representations, the product can be described as follows. Given two representations $\alpha_n : S_n \longrightarrow \mathrm{Aut}(V)$ and $\alpha_m : S_m \longrightarrow \mathrm{Aut}(W)$, $\alpha_n \times \alpha_m : S_n \times S_m \longrightarrow \mathrm{Aut}(V \otimes W)$ and

$\alpha_n \alpha_m : S_{n+m} \longrightarrow \mathrm{Aut}(C[S_{n+m}] \otimes_{S_n \times S_m} (V \otimes W))$.

As an exercise, the reader is invited to show that if χ^2, χ^{21}, χ^{41}, χ^{32}, χ^{221}, χ^{2111}, and χ^{311} denote characters of $S_2, S_3, S_5, S_5, S_5, S_5$, and S_5 respectively (using the notation of the character tables)

then $\chi^2 \chi^{21} = \chi^{41} + \chi^{32} + \chi^{221} + \chi^{311} + \chi^{2111}$

<u>Definition</u>:

$$R(S) = \bigoplus_{n=0}^{\infty} R(S_n)$$

- the sum as abelian groups. The outer products, $R(S_n) \times R(S_m) \to R(S_{n+m})$ for all $n,m \geqslant 0$, induce a multiplication on $R(S)$.

<u>Proposition</u>: $R(S)$ is a graded commutative ring with identity.

<u>Proof</u>: The only non-obvious axioms are associativity and commutativity for multiplication, and the distributive law. Commutativity follows from the fact that $R(S_n \times S_m)$ is isomorphic to $R(S_m \times S_n)$ in the obvious way, and the two subgroups $S_n \times S_m \subset S_{n+m}$ and $S_m \times S_n \subset S_{n+m}$ are conjugate. Distributivity follows from the fact that the operation of inducing representations is additive.

To show associativity, it is enough to show, given elements $\alpha_n \in R(S_n)$, $\alpha_m \in R(S_m)$, and $\alpha_p \in R(S_p)$ that

$$(\alpha_n \alpha_m)\alpha_p = \mathrm{Ind}_{S_n \times S_m \times S_p}^{S_{n+m+p}} (\alpha_n \times \alpha_m \times \alpha_p) = \alpha_n(\alpha_m \alpha_p) .$$

We will just prove the left-hand equality, the other being similar. Writing this proposed equality out in more detail, we get

$$\mathrm{Ind}_{S_{n+m} \times S_p}^{S_{n+m+p}} \left(\mathrm{Ind}_{S_n \times S_m}^{S_{n+m}} (\alpha_n \times \alpha_m) \times \alpha_p \right) = \mathrm{Ind}_{S_n \times S_m \times S_p}^{S_{n+m+p}} (\alpha_n \times \alpha_m \times \alpha_p)$$

Let $\sigma \in S_{n+m+p}$ and apply the formula for induced characters (p. 96) to both sides. The right hand side yields

$$\frac{(n+m+p)!}{n!\, m!\, p!} \; \frac{1}{[\sigma]} \sum_{\substack{\sigma_1 \times \sigma_2 \times \sigma_3 \in S_n \times S_m \times S_p \\ \sigma_1 \times \sigma_2 \times \sigma_3 \sim \sigma \text{ in } S_{n+m+p}}} \left(\alpha_n(\sigma_1)\alpha_m(\sigma_2)\alpha_p(\sigma_3) \right)$$

The left hand side yields

$$\frac{(n+m+p)!}{(n+m)!\, p!} \; \frac{1}{[\sigma]} \sum_{\substack{\sigma_q \times \sigma_3 \in S_{n+m} \times S_p \\ \sigma_q \times \sigma_3 \sim \sigma \text{ in } S_{n+m+p}}} \left(\left(\mathrm{Ind}_{S_n \times S_m}^{S_{n+m}} (\alpha_n \times \alpha_m) \right) \alpha_p(\sigma_3) \right)$$

where $\mathrm{Ind}_{S_n \times S_m}^{S_{n+m}} (\alpha_n \times \alpha_m) = \dfrac{(n+m)!}{n!\, m!} \; \dfrac{1}{[\sigma_q]} \displaystyle\sum_{\substack{\sigma_1 \times \sigma_2 \in S_n \times S_m \\ \sigma_1 \times \sigma_2 \sim \sigma_q \text{ in } S_{n+m}}} \left(\alpha_n(\sigma_1)\alpha_m(\sigma_2) \right)$

Given a typical element $\sigma_1 \times \sigma_2 \times \sigma_3 \in S_n\ S_m\ S_p$, the term $\alpha_n(\sigma_1)\alpha_m(\sigma_2)\alpha_p(\sigma_3)$ is counted <u>once</u> in the right hand side, but on the left side this term is counted once for each $\sigma_q \sim \sigma_1 \times \sigma_2$ in S_{n+m}. Hence the factor $[\sigma_q]$ enters to eliminate the duplication. Thus the two sides are indeed equal, proving associativity. ∎

We now construct the map $\Theta: R(S) \longrightarrow \Lambda$ involved in the fundamental theorem. Let a representation $S_n \longrightarrow \mathrm{Aut}\, W$ of S_n be given, and let V be any vector space. The group S_n then acts on the vector space $W \otimes V \otimes V \otimes \ldots \otimes V$ (n copies of V) by, for $\sigma \in S_n$, $\sigma(w \otimes v_1 \otimes \ldots \otimes v_n) = \sigma(w) \otimes v_{\sigma^{-1}(1)} \otimes v_{\sigma^{-1}(2)} \otimes \ldots \otimes v_{\sigma^{-1}(n)}$. Let $W(V)$ be the subspace of $W \otimes V^{\otimes n}$ of vectors fixed under this action: $W(V) = (W \otimes V^{\otimes n})^{S_n}$.

For example, if W is the trivial one-dimensional representation of S_n, $W(V)$ is the n-fold symmetric power of V. If W is the alternating representation of S_n (the nontrivial one-dimensional representation), $W(V)$ is the exterior power.

The construction $V \longmapsto W(V)$, for fixed W, is functorial in V. Given a map $T: V_1 \longrightarrow V_2$, we have $T^{\otimes n}: V_1^{\otimes n} \longrightarrow V_2^{\otimes n}$ and hence $1_W \otimes T^{n}: W \otimes V_1^{\otimes n} \longrightarrow W \otimes V_2^{\otimes n}$. This map is compatible with the action of S_n, so gives a map

$$1_W \otimes T^{\otimes n}\Big|_{(W \otimes V_1^{\otimes n})^{S_n}} : (W \otimes V_1^{\otimes n})^{S_n} \longrightarrow (W \otimes V_2^{\otimes n})^{S_n}$$

or, as we might write, $W(T): W(V_1) \longrightarrow W(V_2)$.

Suppose that $V_1 = V_2$, and that the transformation T is diagonalizable and has eigenvalues $t_1,\ldots,t_k$, (k=dim V) so that $\mathrm{Trace}(T)=t_1+\ldots+t_k$. The eigenvalues of $T^{\otimes n}$ are n^{th}-degree monomials in $t_1,\ldots,t_k$ and similarly for $1_W \otimes T^{\otimes n}$. Recall the elementary

<u>Lemma</u>: Let $S: U \longrightarrow U$ be a linear transformation. Suppose some subspace $U_1 \subset U$ satisfies $S(U_1) \subset U_1$. Then the eigenvalues of the restriction of S to U_1 are a subset of the eigenvalues of S acting on U. ∎

Hence the eigenvalues of W(T) are monomials of degree n in $t_1,\ldots,t_k$, and so the trace of W(T) is a homogeneous polynomial of degree n, with integer coefficients, in $t_1,\ldots,t_k$. This polynomial is symmetric in the variables $t_1,\ldots,t_k$ for the following reason: any permutation of the symbols $t_1,\ldots,t_k$ corresponds to a change of basis of V. But since the given data and the calculation of W(T) are coordinate-free, the answer would be unaffected by such a coordinate change.

Finally, we claim that the expression of the symmetric functions W(T) in terms of elementary symmetric functions $a_1=t_1+..+t_k,\ldots,a_k=t_1..t_k$ is independent of k=dim V, as long as $k \geq n$. Indeed, suppose $k > n$. Then, still, any monomial $t_1^{i_1} \ldots t_k^{i_k}$ has degree n and so cannot involve nontrivially more than n of the variables $t_1,\ldots,t_k$. Since W(T) is symmetric, such a monomial would occur in W(T) if and only if a monomial with the same pattern of powers $i_1,\ldots,i_k$ occurs in W(T) involving just the variables $t_1,\dots,t_n$. Hence the expression of $W(T)$ in terms of monomial symmetric functions is independent of $k \geq n$, and so likewise for the elementary symmetric functions.

Thus, after all this work, we have a map: given a representation $S_n \longrightarrow W$ in $R(S_n)$, there is an associated symmetric function $\Theta(W) \in \Lambda_n$.

<u>Lemma</u>: $\Theta(W_1 \oplus W_2) = \Theta(W_1) + \Theta(W_2)$

<u>Proof</u>:

$$(W_1 \oplus W_2) \otimes V^{\otimes n} = \left((W_1 \otimes V^{\otimes n}) \oplus (W_2 \otimes V^{\otimes n}) \right) \text{ for any } V.$$

Since S_n acts independently on the two factors

$$(W_1 \otimes V^{\otimes n})^{S_n} \oplus (W_2 \otimes V^{\otimes n})^{S_n} = \left((W_1 \otimes V^{\otimes n}) \oplus (W_2 \otimes V^{\otimes n}) \right)^{S_n} .$$

Hence, for a linear transformation T,

$Tr(W_1 \oplus W_2(T)) = Tr(W_1(T) \oplus W_2(T))$ and this latter is equal to $Tr(W_1(T)) + Tr(W_2(T))$. ∎

<u>Corollary</u>: Θ gives a well-defined map $R(S_n) \longrightarrow \Lambda_n$. ∎

<u>Lemma</u>: Θ is multiplicative. I.e., if $\alpha_{n+m} \in R(S_{n+m})$ is the outer product of $\alpha_n \in R(S_n)$ and $\alpha_m \in R(S_m)$, then $\Theta(\alpha_{n+m}) = \Theta(\alpha_n)\Theta(\alpha_m)$, where the product of symmetric functions is taken in $\Lambda = \bigoplus \Lambda_n$.

<u>Proof</u>: We can assume α_n and α_m are actual representations: $\alpha_n: S_n \longrightarrow \mathrm{Aut}\, W_n$, $\alpha_m: S_m \longrightarrow \mathrm{Aut}\, W_m$, $\alpha_{n+m}: S_{n+m} \longrightarrow \mathrm{Aut}\, W_{n+m}$. Let V be any vector space.

$$W_{n+m} \otimes V^{\otimes n+m} = \mathrm{Ind}\,(W_n \otimes W_m) \otimes V^{\otimes n+m}$$

$$= \mathrm{Ind}\,(W_n \otimes W_m \otimes (\mathrm{Res}^{S_n \times S_m}_{S_{n+m}} V^{\otimes n+m})) \quad \text{(by Frobenius Reciprocity)}$$

$$= \mathrm{Ind}\,(W_n \otimes W_m \otimes V^{\otimes n} \otimes V^{\otimes m})$$

$$= \mathrm{Ind}\,((W_n \otimes V^{\otimes n}) \otimes (W_m \otimes V^{\otimes m})) .$$

Hence

$$(W_{n+m} \otimes V^{\otimes n+m})^{S_{n+m}} = \mathrm{Ind}((W_n \otimes V^{\otimes n}) \otimes (W_m \otimes V^{\otimes m}))^{S_{n+m}}$$

$$= (W_n \otimes V^{\otimes n})^{S_n} \otimes (W_m \otimes V^{\otimes m})^{S_m} \qquad \text{(by p. 121)}$$

This isomorphism $W_{n+m}(V) = W_n(V) \otimes W_m(V)$ is functorial in V, so if T is any linear operator with eigenvalues $t_1, \dots, t_q$, $q \;\; n+m$, it induces an equality $W_{n+m}(T) = W_n(T) \otimes W_m(T)$. Since the trace of a tensor product is the product of the traces, this gives $\Theta(W_{n+m}) = \Theta(W_n)\Theta(W_m)$. ∎

Hence the maps $\Theta: R(S_n) \longrightarrow \Lambda_n$, $n=0,1,2,\dots$, can be added up to give a map of rings $\Theta: R(S) \longrightarrow \Lambda$.

Lemma: $\Theta: R(S_n) \longrightarrow \Lambda_n$ is onto, for each $n \geq 0$.

Proof: One basis for Λ_n is $\{h_\pi \mid \pi \vdash n\}$. Such an $h_\pi = h_{\lambda_1} h_{\lambda_2} \cdots h_{\lambda_n}$ is $\Theta(\mathrm{Ind}_{S_{\lambda_1} \times S_{\lambda_2} \times \cdots \times S_{\lambda_n}}^{S_n} 1) = \Theta$(the outer product of the unit representations of $S_{\lambda_1}, \ldots, S_{\lambda_n}$).

Corollary: $\Theta: R(S_n) \longrightarrow \Lambda_n$ is one-one for all $n \geq 0$.

Proof: An onto homomorphism between two free abelian groups of the same finite rank must also be one-one.

An immediate corollary of the fact that for each n, $R(S_n) = \Lambda_n$, (it is convenient to think of Θ as the identity map), is the fact that the characters of the groups S_n are all integer-valued functions. Indeed, the characters $\Theta^{-1}(h_{\lambda_1} \cdots h_{\lambda_n}) = \mathrm{Ind}_{S_{\lambda_1} \cdots S_{\lambda_n}}^{S_n} 1$ are clearly integer-valued, and they give an integral basis for $R(S_n)$.

Thus we have shown that $\Theta: R(S) \longrightarrow \Lambda$ is an isomorphism of rings. Since Λ is also a λ-ring, the isomorphism Θ induces a corresponding λ-structure on $R(S)$, and taking $R(S)$ with this additional structure, we have the main result:

Theorem: The Fundamental Theorem of the Representation Theory of the Symmetric Group). The map $\Theta: R(S) \longrightarrow \Lambda$ is an isomorphism of λ-rings.

The induced λ-structure on R(S) assigns, for example, to integers $k, n \geq 1$ and $\alpha \in R(S_k)$, an element $h_n(\alpha) \in (R_{nk})$. Classically these operations on Λ, so by extension, on R(S), were referred to as outer plethysm (to be distinguished from the λ-structure of the individual λ-rings, $R(S_n)$, $n \geq 0$, which is called inner plethysm). One problem is to describe the outer plethysm explicitly. It is a reasonable guess that the operation h_n applied to a representation $\alpha: S_k \longrightarrow V$ is performed by first constructing the induced representation of the wreath product $S_n[S_k] \longrightarrow V^{\otimes n}$, and, using the natural inclusion $S_n[S_k] \subset S_{nk}$, inducing to a representation of S_{nk}. But we have been unable to verify this explicitly. Another outstanding problem is to find reasonable algorithms for computing outer (or inner) plethysm. Few calculations have been made (see Littlewood [27], [28]).

The point of the fundamental theorem is that it allows us to pass freely back and forth, for each integer n, between the representation ring $R(S_n)$ (=the ring of characters of $R(S_n)$) on one hand. and the group of symmetric functions of weight n (=the group of λ-operations "of weight n") on the other. On each side of the equality there are a number of calculations one can do, a number of obvious bases for the abelian group involved, and some "canonical" elements, and the game is to relate these. This project is carried out in the rest of this chapter.

2. Complements and Corollaries

Let $n \geq 1$ be a fixed integer and let $\Theta: R(S_n) \longrightarrow \Lambda_n$ be the isomorphism of abelian groups given in the Fundamental Theorem. Given a basis for the group $R(S_n)$, its image under Θ is a basis for Λ_n, and vice-versa. For example, the symmetric functions $\{h_\pi \mid \pi \vdash n\}$ are a basis of Λ_n and we have already noted that, if π is the partition $\pi = (\lambda_1, \ldots, \lambda_n)$ of n,

$$\Theta^{-1}(h_\pi) = \mathrm{Ind}_{S_{\lambda_1} \times \ldots \times S_{\lambda_n}}^{S_n} 1$$

where 1 is the product $1 \times \ldots \times 1$ of the trivial one-dimensional representations of $S_{\lambda_1}, \ldots, S_{\lambda_n}$. (Following the notation of Littlewood [29]) this representation is often denoted φ_π. When it is necessary to distinguish between $R(S_n)$ and Λ_n, we will adhere to this notation: h_π is the symmetric function and φ_π is the associated representation of S_n. But in general it is easier to identify $R(S_n)$ with Λ_n via Θ and use h_π for both objects.

In the case of the basis of Λ_n by products of elementary symmetric functions, $\{a_\pi \mid \pi \vdash n\}$, $\Theta^{-1}(a_\pi)$ is the representation

$$\Theta^{-1}(a_\pi) = \mathrm{Ind}_{S_{\lambda_1} \times \ldots \times S_{\lambda_n}}^{S_n} (\mathrm{alt})$$

where $\pi = (\lambda_1, \ldots, \lambda_n)$, and (alt) is the product of the alternating representations on the groups S_{λ_i}. As above, we will be sloppy and write "a_π" not only for the symmetric function a_π but also for the representation $\Theta^{-1}(a_\pi)$, again treating Θ as an identity map.

So far, only the abelian group structure of $R(S_n)$ has appeared. But $R(S_n)$ is also a λ-ring with a dot product. To avoid confusion with the outer product defined above

$$R(S_n) \times R(S_m) \longrightarrow R(S_{n+m}) \qquad n,m \geq 0$$

$$f,g \rightsquigarrow fg$$

the usual representation-theoretic product

$$R(S_n) \times R(S_n) \longrightarrow R(S_n) \qquad n \geq 0$$

will be called, in this chapter, the <u>inner product</u>, and denoted $f,g \rightsquigarrow f*g$.

The scalar product

$$R(S_n) \times R(S_n) \longrightarrow \mathbb{Z} \qquad n \geq 0$$

will be called, in this chapter, the <u>dot product</u>, and denoted $f,g \rightsquigarrow f.g$.

It is essential to keep these three products straight, and we will adhere rigorously to these terms throughout this chapter.

As we have seen, the Fundamental Theorem corresponds the outer product to the usual product of symmetric functions. But since $\Theta: R(S_n) \longrightarrow \Lambda_n$ is an isomorphism, we can induce an <u>inner product</u> and a <u>dot product</u> on Λ_n from that on $R(S_n)$. A bit later we will give explicit formulas for these. (p. 145)

Using the inner product, the element $a_n \in R(S_n)$ gives a map θ_n of the group $R(S_n)$, $\sigma \rightsquigarrow a_n * \sigma$. Since $a_n * a_n = h_n$, and $h_n * \sigma = \sigma$ for all $\sigma \in R(S_n)$, θ_n is an involution: $\theta_n^2 = 1$. We let the symbol θ_n refer also to the induced involution on Λ_n. Let $\theta : \Lambda \longrightarrow \Lambda$ be the map of graded groups given by $\theta = \bigoplus \theta_n$. θ can also be specified by $\theta(a_\pi) = h_\pi$, all π n, all $n \geq 0$ and again, also by $\theta(h_\pi) = a_\pi$. Another characterization is given on p. 181 . Note θ is <u>not</u> a ring homomorphism.

The main result in this section is the following theorem.

<u>Theorem</u>: The element $L_\pi \in R(S_n)$ defined (as previously) by the character

$$L_\pi(\sigma) = \begin{cases} \dfrac{n!}{[\pi]} & \sigma \in [\pi] \\ 0 & \sigma \notin [\pi] \end{cases}$$

corresponds under Θ to the power sum function s_π .

<u>Proof</u>: First we show the theorem for the trivial partition of n into one part: $\pi = (n)$. Then since Θ is multiplicative, it is only necessary to show that, if $\pi = (\lambda_1, \ldots, \lambda_n)$, L_π is the outer product of $L_{\lambda_1}, \ldots, L_{\lambda_n}$.

Since $\dfrac{n!}{[(n)]} = n$, $L_n(\sigma) = n$ or 0, depending on whether σ is an n-cycle or not. s_n is the symmetric function $\xi_1^n + \xi_2^n + \ldots$ in Λ, and the immediate object is to show that $\Theta(L_n) = s_n$.

This will be accomplished by induction on n. In the case $n=1$, $L_1=\varphi_1$, $\Theta(\varphi_1)=h_1$, and $h_1=s_1$, so the statement is trivial. For $n>1$, we use the Newton formula

$$s_n + s_{n-1}h_1 + s_{n-2}h_2 + \dots + s_1h_{n-1} = nh_n \quad .$$

Suppose the theorem is true for $s_1, s_2, \dots, s_{n-1}$. We will evaluate the character associated to s_n on $R(S_n)$.

First we evaluate $\Theta^{-1}(s_r h_{n-r}) = L_r * \varphi_{n-r}$, $1 \leq r \leq n-1$. Let $\sigma \in S_n$ have cycle structure $(1^\alpha 2^\beta \dots n^\gamma)$.

$$\Theta^{-1}(s_r h_{n-r})(\sigma) = L_r * \varphi_{n-r}(\sigma)$$

$$= \binom{n}{r} \frac{1}{[\sigma]} \sum_{\substack{\sigma_1 \in S_r \times S_{n-r} \\ \sigma_1 \sim \sigma \text{ in } S_n}} (L_r \times 1)(\sigma_1)$$

(applying the formula for induced characters of p.).

Given $\sigma_1 \in S_r \times S_{n-r}$, $(L_r \times 1)(\sigma_1)$ will be zero unless σ_1 is of the form

$$\sigma_1 = (\qquad\qquad) \cdot \underbrace{(\quad) \cdot (\quad) \cdot \dots \cdot (\quad)}$$

an r-cycle containing the numbers $1, \dots, r$ in some order; other cycles involving the numbers from $r+1$ to n

If σ_1 is of this form, $(L_r \times 1)(\sigma_1) = r$.

Hence we want to count the ratio

$$\frac{\left(\begin{array}{l}\text{the number of } \sigma_1 \text{ in } S_r \ \ S_{n-r} \text{ which} \\ \text{are conjugate in } S_n \text{ to } \sigma\end{array}\right)}{\left(\begin{array}{l}\text{the number of } \sigma_1 \text{ in } S_n \text{ which are} \\ \text{conjugate in } S_n \text{ to } \sigma\end{array}\right)}$$

and $\Theta^{-1}(s_r h_{n-r}) = \binom{n}{r} \cdot r \cdot (\text{this ratio})$.

Given that σ has cycle structure $(1^\alpha 2^\beta \ldots r^\eta \ldots)$, the denominator is $[1^\alpha 2^\beta \ldots r^\eta .]$ (using the notation of p.). The numerator of the ratio is then the number of ways of taking one r-cycle on the letters $1,2,\ldots,r$ times the number of permutations of type $(1^\alpha 2^\beta \ldots r^{\eta-1} \ldots)$ on the remaining letters $r+1,\ldots,n$. Hence the numerator is $[r].[1^\alpha 2^\beta \ldots r^{\eta-1} \ldots]$. The ratio is then

$$\frac{\left(\dfrac{r!}{r^1 1!}\right)\left(\dfrac{(n-r)!}{1^\alpha \alpha! 2^\beta \beta! \ldots r^{\eta-1}(\eta-1)! \ldots}\right)}{\left(\dfrac{n!}{1^\alpha \alpha! 2^\beta \beta! \ldots r^\eta \eta! \ldots}\right)} = \frac{\eta}{\binom{n}{r}}$$

Hence $\Theta^{-1}(s_r h_{n-r})(\sigma) = 1_r * \varphi_{n-r}(\sigma) = r\binom{n}{r}\dfrac{\eta}{\binom{n}{r}} = \eta r$ if σ has cycle structure $(1^\alpha 2^\beta \ldots r^\eta \ldots)$.

Note that ηr is just the number of letters from $1,2,\ldots,n$ contained in some r-cycle of σ.

Hence $\Theta^{-1}(s_1h_{n-1}+\ldots+s_{n-1}h_1)(\sigma) = \alpha + 2\beta + \ldots + r\eta + \ldots$
= the number of elements from 1,2,...,n not in an n-cycle of σ.

Applying Newton's formula,

$$\Theta^{-1}(s_n)(\sigma) = \Theta^{-1}(nh_n - s_{n-1}h_1 - \ldots - s_1h_{n-1})(\sigma)$$

$$= nh_n(\sigma) - \Theta^{-1}(s_{n-1}h_1 + \ldots + s_1h_{n-1})(\sigma)$$

= n - (the number of elements from 1,2,...,n not contained in an n-cycle of σ)

= the number of letters in an n-cycle of σ

$$= \begin{cases} n & \text{if } \sigma \text{ is an n-cycle} \\ 0 & \text{if } \sigma \text{ is not an n-cycle} \end{cases}$$

$$= L_n(\sigma).$$

Hence

$$\Theta(L_n) = s_n \quad \text{all } n \geq 1.$$

Now we must prove that, for partitions $\pi_1=(1^{\alpha_1}2^{\beta_1}\ldots)$ of n_1 $\pi_2=(1^{\alpha_2}2^{\beta_2}\ldots)$ of n_2, the outer product of L_{π_1} and L_{π_2} is $L_{\pi_1\pi_2}$ where $\pi_1\pi_2=(1^{\alpha_1+\alpha_2}2^{\beta_1+\beta_2}\ldots)$ is a partition of n_1+n_2. The theorem will then follow by induction on the number of parts of n.

Given $\sigma \in S_{n_1+n_2}$, with cycle structure $\pi=(1^{\alpha}2^{\beta})$, we calculate

$\mathrm{Ind}_{S_{n_1}\times S_{n_2}}^{S_n}(L_{\pi_1}\times L_{\pi_2})(\sigma)$ using the formula for induced characters:

$$\mathrm{Ind}_{S_{n_1}\times S_{n_2}}^{S_n}(L_{\pi_1}\times L_{\pi_2})(\sigma)=\frac{(n_1+n_2)!}{n_1!n_2!}\,\frac{1}{[\pi]}\sum_{\substack{(\sigma_1\times\sigma_2)\in S_{n_1}\times S_{n_2}\\ \sigma_1\times\sigma_2\sim\sigma \text{ in } S_n}}\Big(L_{\pi_1}(\sigma_1)L_{\pi_2}(\sigma_2)\Big)$$

$(n=n_1+n_2)$

Observe that

$$L_{\pi_1}(\sigma_1)\,L_{\pi_2}(\sigma_2)=\begin{cases}\dfrac{n_1!}{[\pi_1]}\,\dfrac{n_2!}{[\pi_2]} & \text{if each } \sigma_i \text{ has cycle structure } (1^{\alpha_i}2^{\beta_i}\ldots) \text{ - so necessarily } \sigma \text{ has cycle structure } \pi_1\pi_2 = (1^{\alpha_1+\alpha_2}2^{\beta_1+\beta_2}\ldots)\\ 0 & \text{otherwise}\end{cases}$$

Hence

$$\mathrm{Ind}_{S_{n_1}\times S_{n_2}}^{S_n}L_{\pi_1}\times L_{\pi_2}(\sigma)=\begin{cases}0 & \pi\neq\pi_1\pi_2\\ \dfrac{(n_1+n_2)!}{n_1!\,n_2!}\,\dfrac{1}{[\pi_1\pi_2]}\,\dfrac{n_1!}{[\pi_1]}\,\dfrac{n_2!}{[\pi_2]}\cdot N & \pi=\pi_1\pi_2\end{cases}$$

where N is the number of elements $(\sigma_1\times\sigma_2)$ of $S_{n_1}\times S_{n_2}$ of cycle structure $\pi_1\pi_2$. This number is $[\pi_1][\pi_2]$. Hence

$$\mathrm{Ind}_{S_{n_1}\times S_{n_2}}^{S_n}L_{\pi_1}\times L_{\pi_2}(\sigma)=\left\{\begin{matrix}0 & \text{if } \pi\neq\pi_1\pi_2\\ \dfrac{(n_1+n_2)!}{[\pi_1\pi_2]} & \text{if } \pi=\pi_1\pi_2\end{matrix}\right\}=L_{\pi_1\pi_2}(\sigma)$$

Hence $L_{\pi_1\pi_2}$ is the outer product of L_{π_1} and L_{π_2}. ∎

We have so far four properties of the isomorphism $\Theta: R(S_n) \longrightarrow \Lambda_n$, any one of which characterize it uniquely among all group homomorphisms $R(S_n) \longrightarrow \Lambda_n$:

1) The original definition

2) $\Theta^{-1}(h_\pi) = \mathrm{Ind}_{S_\pi}^{S_n} 1$ all $\pi \vdash n$

3) $\Theta^{-1}(a_\pi) = \mathrm{Ind}_{S_\pi}^{S_n} \underline{\mathrm{alt}}$ all $\pi \vdash n$

4) $\Theta(L_\pi) = s_\pi$ all $\pi \vdash n$.

The next few corollaries will give several more ways to define this map.

First is a way of defining Θ which goes back to Frobenious and is mentioned in Weyl's <u>Classical Groups</u> ([45], p.215). First recall the definition of the central functions K_π on S_n: for $\sigma \in S_n$, $K_\pi(\sigma)$ is 1 or 0 according as to whether σ is in the conjugacy class π or not. Thus $L_\pi = (n!/[\pi])K_\pi$. Given now any character χ of S_n, $\chi = \sum_{\pi \vdash n} \chi(\pi)K_\pi$, so $\chi = \sum_{\pi \vdash n} ([\pi]/n!)\chi(\pi)L_\pi$.

Hence, using the theorem above, and the fact that Θ is additive, $\Theta(\chi) = \sum_{\pi \vdash n} \chi(\pi) \frac{[\pi]}{n!} s_\pi$ - clearly a quite explicit recipe for the map Θ.

This leads to the formula, promised above, for the calculation of the dot product and inner product in Λ_n. Recall, the dot product $L_{\pi_1} . L_{\pi_2}$ (respectively, the inner product $L_{\pi_1} L_{\pi_2}$) is equal to $(n!/[\pi_1])$ or zero (respectively $(n!/[\pi_1])L_{\pi_1}$ or zero) according as to whether $\pi_1=\pi_2$ or not. Hence

$$s_{\pi_1} \cdot s_{\pi_2} = \begin{cases} \frac{n!}{[\pi_1]} & \text{if } \pi_1=\pi_2 \\ 0 & \text{if } \pi_1 \neq \pi_2 \end{cases}$$

$$s_{\pi_1} * s_{\pi_2} = \begin{cases} \frac{n!}{[\pi_1]}\, s_{\pi_1} & \text{if } \pi_1=\pi_2 \\ 0 & \text{if } \pi_1 \neq \pi_2 \end{cases}.$$

These definitions can be found in some of the fairly ancient literature on the subject. Redfield ([35]) in 1927 defined these using the symbols f ℧ g for the inner product of two symmetric functions f and g, and f Ω g for the dot product.

Another characterization of Θ is the following.

<u>Proposition</u>: Let G be a subgroup of S_n. (In other words, let G be a faithful permutation group of degree n.) Let $\chi = \mathrm{Ind}_G^{S_n} 1$. Then

$$\Theta(\chi) = \frac{1}{|G|} \sum_{\substack{x \in G \\ x = (1^{\alpha} 2^{\beta} \ldots) \\ \text{all } \alpha, \beta, \ldots}} (s_1^{\alpha} s_2^{\beta} \ldots)$$

<u>Proof</u>: We must check that

$$\chi = \frac{1}{|G|} \sum_{\substack{x \in G \\ x \text{ of cycle} \\ \text{structure } \pi}} L_{\pi}$$

and this follows easily from the formula for induced characters. ∎

This construction of a polynomial in the variables $s_1, s_2, \ldots$ associated to a given faithful permutation group is of course, just the classical <u>cycle index</u> of Polya ([33]), (also called the <u>group reduction formula</u> by Redfield ([35]) and Littlewood ([29])).

A worthwhile example (of Redfield) in this context is the pair of subgroups of S_6:

$$G_1 = \begin{cases} (1)(2)(3)(4)(5)(6) \\ (1)(2)(34)(56) \\ (12)(3)(4)(56) \\ (12)(34)(5)(6) \end{cases} \qquad G_2 = \begin{cases} (1)(2)(3)(4)(5)(6) \\ (1)(2)(34)(56) \\ (1)(2)(36)(45) \\ (1)(2)(46)(35) \end{cases}$$

These two subgroups have the same cycle index, namely $\dfrac{s_1^6 + 3s_1^2 s_2^2}{4}$ but are not conjugate as subgroups of S_6. There are also examples of non-isomorphic subgroups of some S_n which yield the same cycle index.

Another classical instance of the map Θ is in the theory of immanents. (See Littlewood [29]). Let there be given an $n \times n$ matrix A:

$$A = \begin{pmatrix} A_{11} & \cdots & A_{1n} \\ \vdots & & \vdots \\ A_{n1} & \cdots & A_{nn} \end{pmatrix}$$

with entries in a given ring.

Let $\sigma \in S_n$ be a permutation. Write

$$P_\sigma = A_{1\sigma(1)} \cdots\cdots A_{n\sigma(n)}$$

Let $\chi \in R(S_n)$ be a character. The <u>immanent of A associated to χ</u>, written $|A|^\chi$ is defined by

$$|A|^\chi = \sum_{\sigma \in S_n} \chi(\sigma) P_\sigma \quad .$$

Thus if χ is the alternating character, we get $|A|^\chi$ = det A, the usual determinant. If χ is the trivial one-dimensional representation of S_n, $|A|^\chi$ is the permanant of A. Etc. If n=3 and χ is the unique 2-dimensional irreducible representation of S_3,

$$|A|^\chi = 2A_{11}A_{22}A_{33} - A_{12}A_{23}A_{31} - A_{13}A_{21}A_{32} \ .$$

In general, for matrices A and B, $|AB|^\chi \neq |BA|^\chi$, if χ is not the alternating representation. This formula does hold, however if A or B is a permutation matrix.

The relevance of immanents to our subject is the following theorem, where the s_i denote the power sum symmetric functions.

Theorem: For $\chi \in R(S_n)$

$$\Theta(\chi) = \frac{1}{n!}\begin{vmatrix} s_1 & 1 & 0 & 0 & \cdots & 0 \\ s_2 & s_1 & 2 & 0 & \cdots & 0 \\ \cdot & \cdot & \cdot & & & \cdot \\ \cdot & \cdot & \cdot & & & \cdot \\ s_{n-1} & s_{n-2} & \cdots & \cdots & s_1 & n-1 \\ s_n & s_{n-1} & \cdots & \cdots & \cdots & s_1 \end{vmatrix}^{\chi}$$

Proof: Both sides are additive in χ, so it is sufficient to prove the identity when χ is one of the characters L_π, for a given partition π. In this case

$$\frac{1}{n!}\ |A|^{\chi} = \sum_{\sigma} \frac{1}{n!} L_\pi(\sigma) P_\sigma = \frac{1}{[\pi]} \sum_{\sigma\in[\pi]} P_\sigma$$

Suppose $\pi = (n_1, n_2, \ldots, n_l)$. (Here the n_i are just the parts of n under π, in no particular order). Suppose $\sigma \in S_n$ is the element $(1,2,\ldots,n_1)(n_1+1,\ldots,n_2)(\ldots$ Then

$$P_\sigma = 1.2.3\ldots(n_1-1).s_{n_1}.(n_1+1)\ldots(n_1+n_2-1).s_{n_2}\ldots\ldots$$

$$= \frac{n!}{n_1(n_1+n_2)(n_1+n_2+n_3)(\ldots\ldots}\ (s_{n_1}.s_{n_2}\ldots.)$$

If σ is any element in the class π, there will be a nonzero P_σ only if no $A_{j,j+2}, A_{j,j+3}, \ldots$ appears. Hence every other nonzero term in the immanent comes from permuting $n_1, n_2, \ldots, n_l$.

If $n_1, n_2, \ldots, n_\ell$ are all different, then

$$\sum_{\sigma \in \pi} P_\sigma = \sum_{\substack{\text{all permutations} \\ n_{i_1}, n_{i_2}, \ldots, n_{i_\ell} \\ \text{of } n_1, n_2, \ldots, n_\ell}} \frac{n!}{(n_{i_1})(n_{i_1}+n_{i_2})(\ \ldots} s_\pi \quad .$$

For the general case when the n_i aren't distinct, we need a

Lemma:

$$\sum_{\substack{\text{all permutations} \\ n_{i_1}, n_{i_2}, \ldots, n_{i_\ell} \\ \text{of } n_1, n_2, \ldots, n_\ell}} \frac{1}{n_{i_1}(n_{i_1}+n_{i_2})(\ldots)(n_{i_1}+\ldots+n_{i_\ell})} = \frac{1}{n_1 n_2 \ldots n_\ell}$$

Proof of Lemma: By induction on ℓ. For $\ell=1$, it is trivial. Suppose it is true for all integers less than ℓ. Then the left hand side is

$$\sum_i \left(\sum_{\substack{\text{all permutations} \\ n_{i_1}, \ldots, n_{i_\ell} \text{ of} \\ n_1, \ldots, n_\ell \text{ with } n_i \\ \text{appearing last: } i_\ell = i}} \frac{1}{(n_{i_1})(n_{i_1}+n_{i_2})\ldots(n_{i_1}+\ldots+n_{i_\ell})} \right)$$

$$= \sum_i \frac{1}{n_1 n_2 \ldots n_{i-1} n_{i+1} \ldots n_\ell} \quad \frac{1}{n_1+n_2+\ldots+n_\ell}$$

(pulling the common last factor from each term and using the induction hypothesis.)

$$= \frac{1}{n_1 n_2 \ldots n_\ell} \sum_i \frac{n_i}{(n_1+n_2+\ldots+n_\ell)} = \frac{1}{n_1 n_2 \ldots n_\ell}$$

\- end of proof of lemma. ∎

Hence if $n_1,\ldots,n_\ell$ are all different, $\sum_{\sigma\in\pi} P_\sigma = \frac{n!}{n_1 n_2 \cdots n_\ell} s_\pi$. In general, though, if $(n_1,n_2,\ldots,n_\ell) = (1^\alpha 2^\beta \ldots)$, $\alpha+\beta+\ldots=\ell$, all the $\ell!$ permutations are not distinct-there are only $(\ell!/\alpha!\beta!\ldots)$ distinct permutations - so the coefficient in general is not

$$\frac{n!}{n_1 n_2 \cdots n_\ell} = \frac{n!}{1^\alpha 2^\beta \cdots} \quad \text{but rather} \quad \frac{n!}{1^\alpha 2^\beta \cdots \alpha!\beta!\cdots} = [\pi] .$$

So $\sum_{\sigma\in\pi} P_\sigma = \frac{n!}{[\pi]} s_\pi$.

Hence

$$\frac{1}{n!}\begin{vmatrix} s_1 & \cdots & 0 \\ \vdots & & \vdots \\ s_n & \cdots & s_1 \end{vmatrix}^{L_\pi} = \frac{1}{[\pi]}\sum P_\sigma = \frac{[\pi]}{[\pi]} s_\pi = s_\pi = \Theta(L_\pi) . \blacksquare$$

This is the classical proof of this fact (taken from Littlewood [29]), but there may be an easier proof. We know to begin with that both sides of $\frac{1}{n!}\left|\vdots\right|^{\chi} = \Theta(\chi)$ are additive in χ, and that the theorem is true for $\chi=a_n$, $n=0,1,2,\ldots$ (p. 44). Hence it would be sufficient to show also that both sides are multiplicative in χ. This we know for $\Theta(\chi)$. But for the left hand side, the matter seems a little more subtle - but of course true since the theorem is true! We leave it to the reader to work out this alternate proof.

Fix the integer n and consider the irreps (irreducible representations) of S_n. By any of the above methods we can write down the associated symmetric functions in Λ_n. A non-obvious fact - the next section is mainly devoted to its proof - is that the resulting functions are the Schur functions $\{ \{\pi\} \mid \pi \vdash n \}$ discussed in Chapter 1. (There we showed that $\{1^n\}=a_n$ and $\{n\}=h_n$ so this takes care of the symmetric functions associated to the one-dimensional irreps.) We will here assume this fact for the rest of this section, mainly for the following notational convenience: since the Schur functions in Λ_n are naturally labeled by partitions of n, so must the irreps of S_n also be so labeled. We will write χ^π for the irrep associated to $\{\pi\}$. Thus χ^{1^n} = <u>alt</u> , and $\chi^n = 1$.

It is interesting to note that this labeling gives a natural one-one correspondence between the set of irreps of S_n and the set of conjugacy classes of S_n: $\chi^\pi \leftrightsquigarrow \{\sigma \mid \sigma \text{ of type } \pi\}$. Of course for any group G, we have shown (p. 84) that these two sets have the same number of elements, but usually there is no such natural correspondence.

Given π a partition of n, we can associate to π the degree of the irrep χ^π, which we will denote H_π. This positive integer is called the "hook number" of π, because of a simple algorithm (p.173) for its computation. From the elementary theory of group representations, one can write such simple identities as

$$H_{1^n} = 1$$

$$H_n = 1$$

$$\sum_{\pi \vdash n} H_\pi^2 = n!$$

A slight bit more involved is the following identity in Λ_n:

<u>Lemma</u>: $$(a_1)^n = \sum_{\pi \vdash n} H_\pi \cdot \{\pi\}$$

<u>Proof</u>: Let U be any representation of a finite group G. Then

$$\bigoplus_{U_i \in \mathrm{Irrep}\, G} (U_i \otimes \mathrm{Hom}_G(U_i, U)) \cong U$$

Here $\mathrm{Hom}_G(U_i, U)$ is as in p. 61. The map $U_i \otimes \mathrm{Hom}_G(U_i, U) \longrightarrow U$ is the obvious one: $u \otimes f \rightsquigarrow f(u)$ and the isomorphism follows since each side is additive in U, and the identity is true when U is irreducible (i.e., when $U = U_i$, some i). Indeed, this is just a restatement of Schur's lemma.

Let now V be a vector space and $U = V^{\otimes n}$. Let $G = S_n$ act on $V^{\otimes n}$ by permutation of factors. Then this identity becomes

$$V^{\otimes n} \cong \bigoplus_{\pi \vdash n} (W_\pi \otimes \mathrm{Hom}_{S_n}(W_\pi, V^{\otimes n}))$$

But $\mathrm{Hom}_{S_n}(W_\pi, V^{\otimes n}) = (\mathrm{Hom}(W_\pi, V^{\otimes n}))^{S_n} = (W_\pi^{\mathrm{dual}} \otimes V^{\otimes n})^{S_n} = W_\pi^{\mathrm{dual}}(V)$

the last equality being the definition of the operation $W_\pi^{\mathrm{dual}}(\ )$ associated to a representation W_π^{dual}.

Now note two things: W_π is an irrep if and only if W_π^{dual} is, and they both have the same degree. (This is true for any group. An irrelevant fact is that for S_n, the irreps W_π and W_π^{dual} are isomorphic.)

Hence:

$$V^{\otimes n} = \bigoplus_{\pi \vdash n} W_\pi^{dual} \otimes W_\pi(V) .$$

This is a functorial isomorphism in V, so replacing V by an endomorphism of V with eigenvalues $t_1, t_2, .., t_n$, and taking the trace of both sides we get

$$(t_1+t_2+\dots)^n = \sum_{\pi \vdash n} H_\pi . \{\pi\}(t_1,\dots,t_n)$$ ∎

Schur functions interpreted as operations give a generalization of binomial coefficients. Namely, let π be any partition of any integer n and let V be a vector space of dimension m.

Write

$$\binom{m}{\pi} = \text{dimension } \{\pi'\}(V)$$

where π' is the partition conjugate to π.

If $\pi=(n)$, $\binom{m}{\pi} = \binom{m}{(n)} = \binom{m}{n} = \dim \{1^n\}(V) = \dim \Lambda^n V$ - the usual binomial coefficient.

The last proposition gives an identity for these generalized binomial coefficients

$$m^n = \sum_{\pi \vdash n} H_\pi \binom{m}{\pi} .$$

In terms of λ-rings, $\binom{m}{\pi}$ is the result of applying the operations $\{\pi'\} \in \Lambda$ to the element m in the (binomial) λ-ring $\mathbb{Z}$.

We close this section by reviewing what we now have. For each integer $n \geq 0$, we have an isomorphism $R(S_n) \xrightarrow{\Theta} \Lambda_n$ which is natural (i.e., extends to a ring isomorphism $R(S) \longrightarrow \Lambda$). We identify these two sets. The resulting object is a free abelian group of rank $\pi(n)$, the number of partitions of n, with an inner product, a dot product, and an involution (inner multiplying by a_n). There are several natural bases (as free abelian group): $\{a_\pi, \pi \vdash n\}$, $\{h_\pi, \pi \vdash n\}$, $\{\langle\pi\rangle, \pi \vdash n\}$, $\{\{\pi\}, \pi \vdash n\}$ and a rational basis $\{s_\pi, \pi \vdash n\}$. Given any two of these bases (or one basis and the set of s_π) one can ask for the transition matrix between them. This is an invertible integer matrix (or $1/n!$ times an integer matrix if the s_π are chosen) and a large part of the combinatorial theory of the symmetric group consists of methods for calculating these matrices explicitly and interpretations of the resulting numbers. In the next two sections we explore some of this theory.

3. Schur Functions and the Frobenius Character Formula

The main object of this section is to prove that the Schur functions are the images under the isomorphism Θ of the irreducible representations of the symmetric groups. It turns out that the method of proof gives a set of formulas which will be basic in the computation of the combinatorial theory of the symmetric group.

The proof comes from the simple observation (pointed out in this context by Philip Hall [20]) that in a free abelian group of finite rank, with a dot product, an orthonormal basis must be unique (if it exists at all) - at least, unique up to sign and order. This comes from the fact that if two orthonormal bases are given, the transition matrix from one to the other must be orthogonal with integer entries. The only orthogonal integral matrices are signed permutation matrices - i.e., square matrices in which each row and each column contain exactly one non-zero entry, that entry being $\pm$ 1.

$R(S_n) = \Lambda_n$ has a dot product, and an orthonormal basis, the irreducible representations. Hence any orthonormal basis in Λ_n must correspond up to sign and order with the irreps. The key lemma is the following.

Lemma: Given any expression $h_n(xy) = \sum_{\lambda \vdash n} q_\lambda(x)\, r_\lambda(y)$ with $q_\lambda, r_\lambda \in \Lambda_n$ (notation as in p. 37), the sets $\{q_\lambda \mid \lambda \vdash n\}$ and $\{r_\lambda \mid \lambda \vdash n\}$ are both bases of Λ_n and are dual to each other under the dot product in Λ_n.

The proof of this lemma requires us first to recall some facts about dot products in free abelian groups.

Let F be a free abelian group of finite rank k with a dot product. Thus, there is a function $(-,-)$ defined on $F \times F$ satisfying, for all $x,y,z \in F$

i) $(x,x) \in \mathbb{Z}$

ii) $(x,x) \geq 0$ and equality holds only if $x=0$

iii) $(x,y) = (y,x)$

iv) $(x,y+z) = (x,y) + (x,z)$

$(x+y,z) = (x,z) + (y,z)$

This gives a map $\beta: F \longrightarrow \mathrm{Hom}(F,\mathbb{Z})$, $x \rightsquigarrow (x,-)$ which is necessarily one-one.

Suppose, now, that the dot product is also normal in the sense that, if $\{r_\lambda\}$ is any basis of F, the $k \times k$ determinant $\det((r_\lambda, r_{\lambda'})) = \pm 1$. (This is true in $F = R(S_n) = \Lambda_n$ since we know that $R(S_n)$ has an orthonormal basis, the irreducible representations.) An equivalent definition of normality is that the map $\beta: F \longrightarrow \mathrm{Hom}(F,\mathbb{Z})$ given above, is onto (hence an isomorphism).

Another equivalent statement is that, given a basis $\{r_\lambda\}$ of F, there is a unique dual basis $\{s_\lambda\}$ of F , dual in the sense that $(r_i,s_j) = \delta_{ij}$ (Kroneckor delta).

Let us prove that normality implies the existence of a dual to a given basis $\{r_\lambda\}$. To find the s_λ, one must solve equations like $s_\lambda = a_{11}r_{\lambda_1}+a_{12}r_{\lambda_2}+\dots$, for <u>integers</u> a_{ij}. Dotting this equation in succession by $r_{\lambda_1}, r_{\lambda_2}, \dots$ we get

$$
\begin{aligned}
1 &= a_{11}(r_{\lambda_1},r_{\lambda_1}) + a_{12}(r_{\lambda_1},r_{\lambda_2}) + \dots \\
\cdot\ & \quad \cdot \\
\cdot\ & \quad \cdot \\
\cdot\ & \quad \cdot \\
0 &= a_{11}(r_{\lambda_i},r_{\lambda_1}) + a_{12}(r_{\lambda_i},r_{\lambda_2}) + \dots \\
\cdot\ & \\
\cdot\ &
\end{aligned}
$$

By Cramer's rule, this set of equations is solvable over $\mathbb{Z}$ since $\det((r_{\lambda_i},r_{\lambda_j}))= \pm 1$ is invertible in $\mathbb{Z}$.

Specifically now in the case of $F=\Lambda_n$, we want to look at the inverse map $\beta^{-1}:\mathrm{Hom}(F,\mathbb{Z}) \longrightarrow F$.

There is a natural isomorphism $\mathrm{Hom}(\mathrm{Hom}(F,\mathbb{Z}),F)= F\otimes F$. Indeed, for any free abelian group of finite rank F, and any group G, there is a natural isomorphism $\mathrm{Hom}(\mathrm{Hom}(F,\mathbb{Z}),G) = F\otimes G$. This map is given by

$$\psi: F \otimes G \longrightarrow \mathrm{Hom}(\mathrm{Hom}(F, \mathbb{Z}), G)$$

by, for

$$\sum f_i \otimes g_i \in F \otimes G \qquad \text{and } \eta \in \mathrm{Hom}(F, \mathbb{Z})$$

$$\psi(\sum f_i \otimes g_i)(\eta) = \sum \eta(f_i) g_i \in G \qquad .$$

Both sides are additive in F and for $F = \mathbb{Z}$, this is easily shown to be an isomorphism. Hence it is an isomorphism in general.

Thus $\beta^{-1} \in \Lambda_n \otimes \Lambda_n$. It is now simply a matter of linear algebra to show that any expansion $\beta^{-1} = \sum_{\lambda \vdash n} q_\lambda \otimes r_\lambda$, $q_\lambda, r_\lambda \in \Lambda_n$ must entail that $\{q_\lambda \mid \lambda \vdash n\}$ and $\{r_\lambda \mid \lambda \vdash n\}$ are both bases and are dual to one another. For, chasing the isomorphism back, $\beta^{-1} = \sum q_\lambda \otimes r_\lambda$, so for $q \in \mathrm{Hom}(\Lambda_n, \mathbb{Z})$, $\beta^{-1}(q) = (\sum r_\lambda \otimes s_\lambda)(q) =$ $= \sum_\lambda q(r_\lambda) s_\lambda \in \Lambda_n$. β^{-1} is onto so this shows that the s_λ must span Λ_n. Since there are just enough of them, they form a basis. Furthermore, for $r \in \Lambda_n$, the definition of β gives $q = \beta(r) =$ the map $\Lambda_n \longrightarrow \mathbb{Z}$ given by dotting with r: $q = (r, -)$. Thus $r = \beta^{-1}\beta(r) = \beta^{-1}(q) = \sum_\lambda (r, r_\lambda) s_\lambda$. Let $r = s_{\lambda'}$. Then

$s_{\lambda'} = \sum_\lambda (s_{\lambda'}, r_\lambda) s_\lambda$. Since the s_λ are a basis, $(s_{\lambda'}, r_\lambda) = 1$ or 0, according as to whether $\lambda' = \lambda$ or not. Hence the set of r_λ is a dual basis to the set of s_λ.

This argument is reversible: given any basis q_λ, and dual basis r_λ, of Λ_n, β^{-1} can be expanded as $\sum_{\lambda \vdash n} q_\lambda \otimes r_\lambda$.

To identify $\beta^{-1} \in \Lambda_n \otimes \Lambda_n$, it would be sufficient to tensor everything with the rational numbers ("introduce denominators") and find the element β^{-1} in $(\Lambda_n \otimes \mathbb{Q}) \otimes_{\mathbb{Q}} (\Lambda_n \otimes \mathbb{Q})$. Any expression $\beta^{-1} = \sum_{\lambda \vdash n} q_\lambda \otimes r_\lambda$ with q_λ, r_λ not only in $\Lambda_n \otimes \mathbb{Q}$ but in $\Lambda_n \subset \Lambda_n \otimes \mathbb{Q}$ would still give dual bases of Λ_n.

Now the argument is brought back from generality to the case at hand by noticing that we have the expansion

$$h_n(xy) = \sum_{\lambda \vdash n} \frac{[\lambda]}{n!} s_\lambda(x)\, s_\lambda(y) \qquad \text{(p. 39)}$$

and, $\{([\lambda]/n!)s_\lambda \mid \lambda \vdash n\}$ and $\{s_\lambda\}$ are dual bases of $\Lambda_n \otimes \mathbb{Q}$ (p. 145). Hence $h_n(xy) = \beta^{-1}$. ■

Corollary: For all $n \geq 1$

i) $\{<\lambda> \mid \lambda \vdash n\}$ and $\{h_\lambda \mid \lambda \vdash n\}$ are dual bases of Λ_n.

ii) $\{\{\lambda\} \mid \lambda \vdash n\}$ is an orthonormal basis of Λ_n.

iii) Let χ be an irreducible representation of S_n. Then $\Theta(\chi) \in \Lambda_n$ is of the form $\Theta(\chi) = \pm \{\lambda\}$ for some $\lambda \vdash n$. ∎

Proof: See pp. 38, 43

Later (p. 169) we will show that the $\pm$ sign in iii) is in fact always a plus sign. Hence we can write, for each $\lambda \vdash n$, χ^λ for $\Theta^{-1}(\{\lambda\})$. We will also prove that, if λ' is the partition conjugate to λ, then $\chi^{\lambda'} = a_n * \chi^\lambda$. In other words, the endomorphism on Λ (p. 139) comes from taking the basis of Λ by Schur functions and mapping $\{\lambda\} \rightsquigarrow \{\lambda'\}$.

Another consequence of the correspondence between Schur functions and irreps of S_n is a number of formulas for Schur functions. These are deduced by using the various definitions given in the last section for the map Θ. Thus

$$\{\lambda\} = \frac{1}{n!} \begin{vmatrix} s_1 & 1 & 0 & \cdots\cdots \\ s_2 & s_1 & 2 & 0 \quad \cdots \\ \cdot\ \cdot & & & \\ s_n & s_{n-1} & \cdot\ \cdot & s_1 \end{vmatrix}^{\chi^\lambda}$$

(using the definition of Θ by immanents)

and

$$\{\lambda\} = \sum_{\mu} \frac{\chi^{\lambda}(\mu)\ [\mu]}{n!}\ s_{\mu}$$

$$s_{\mu} = \sum_{\lambda} \chi^{\lambda}(\mu)\ \{\lambda\}$$

(using the facts that $\Theta(L_{\mu})=s_{\mu}$, $\Theta(\chi^{\lambda}) = \{\lambda\}$, and that these formulas hold (p. 90) between the χ^{λ} and the L_{μ}.)

The most important result from the identification of $h_n(xy) \in \Lambda_n \otimes \Lambda_n$ comes from the fact that we have three different expansions of $h_n(xy)$ in the form $\sum_{\lambda \vdash n} q_{\lambda} \otimes r_{\lambda}$, $\{q_{\lambda}\}, \{r_{\lambda}\}$ dual bases. Each pair of expansions gives rise to formulae relating the symmetric functions involved.

The easiest derivation comes from the identity in $\Lambda_n \otimes \Lambda_n$: $h_n(xy) = \sum_{\pi \vdash n} \frac{n!}{[\pi]} s_{\pi}(x) s_{\pi}(y) = \sum_{\pi \vdash n} h_{\pi}(x) \langle\pi\rangle(y)$. Identifying $\Lambda_n(x)$ with $R(S_n)$, we get

$$\sum_{\pi \vdash n} \frac{n!}{[\pi]} L_{\pi} s_{\pi}(y) = \sum_{\pi \vdash n} \varphi_{\pi} \langle\pi\rangle(y)$$

where $\varphi_{\pi} = \Theta^{-1}(h_{\pi}) = \mathrm{Ind}_{S_{\pi}}^{S_n} 1$, as usual. Both sides are now in $R(S_n) \otimes \Lambda_n$. $R(S_n)$ being a ring of functions, class functions on S_n, both sides can be applied to the class ρ of S_n. The result is

$$s_\rho(y) = \sum_{\pi \vdash n} \varphi_\pi(\rho) \langle\pi\rangle(y)$$

Hence the characters $\varphi_\pi(\rho)$ give the transition matrix between the basis $\{\langle\pi\rangle \mid \pi\vdash n\}$ of Λ_n and the rational basis $\{ s_\rho \mid \rho\vdash n\}$. Conversely, since the symmetric functions s_ρ can easily be expressed in terms of monomial symmetric functions, this gives an algorithm for computing the characters φ_π.

Similarly, identifying $\Lambda_n(y)$ with $R(S_n)$, we have

$$s_\rho(x) = \sum_{\pi\vdash n} \langle\pi\rangle(\rho)\ h_\pi(x)$$

showing that the characters of S_n corresponding to the monomial symmetric functions give the transition matrix between the s_ρ amd the h_π.

It is worthwhile here to mention another kind of symmetric functions which have not been much studied (except quite recently - see [13]). Let us call these $\{f_\pi \mid \pi\vdash n\}$, the "forgotten symmetric functions". They are defined as the dual basis to $\{a_\pi \mid \pi\vdash n\}$. Equivalently, $f_\pi = a_n * \langle\pi\rangle$ where $\pi\vdash n$, $n=0,1,2,\ldots$ As above, one deduces

$$h_n(xy) = \sum_{\lambda\vdash n} f_\lambda(x)\ a_\lambda(y)$$

$$s_\rho(x) = \sum_{\pi\vdash n} f_\pi(\rho) a_\pi(x)$$

and $$s_\rho(x) = \sum_{\pi \vdash n} a_\pi(\rho)\, f_\pi(x)$$

(where in the second equation, we have denoted by f_π, the character associated to the symmetric function f_π).

The next pair of expressions for $h_n(xy)$ gives the identity

$$h_n(xy) = \sum_{\pi \vdash n} \frac{n!}{[\pi]}\, s_\pi(x)\, s_\pi(y) = \sum_{\lambda \vdash n} \{\lambda\}(x)\{\lambda\}(y) \quad .$$

Identifying $\Lambda_n(y) = R(S_n)$ as above, and evaluating on the class ρ, we get the

<u>Frobenius Character Formula</u>

$$s_\rho(x) = \sum_{\lambda \vdash n} \chi^\lambda(\rho)\ \{\lambda\}\ (x)$$

It is worthwhile to write this out more explicitly. Take n variables $x_1, x_2, .., x_n$. Recall that, for an integer m, $s_m(x) = x_1^m + x_2^m + .. + x_n^m$. For a partition $\rho = (\rho_1, \rho_2, \ldots)$, $s_\rho = s_{\rho_1} s_{\rho_2} \ldots$, again a symmetric function in the x_i's. The definition of Schur function was, for $\lambda = (\lambda_1, \lambda_2, .., \lambda_n)$, $\lambda_1 \geq \lambda_2 \geq \ldots \geq \lambda_n \geq 0$, a partition

$$\{\lambda\} = \frac{\displaystyle\sum_{\sigma \in S_n} \operatorname{sgn} \sigma \left(x_{\sigma(1)}^{\lambda_1+n-1}\, x_{\sigma(2)}^{\lambda_2+n-2} \ldots x_{\sigma(n)}^{\lambda_n} \right)}{\Delta(x_1, x_2, \ldots, x_n)}$$

where $\Delta(x_1, x_2, .., x_n) = \sum_{i<j} (x_i - x_j) \quad .$

Thus we get the classical expression of the Frobenius Character Formula

$$s_\rho(x)\ \Delta(x_1,\ldots,x_n) = \sum_{\lambda\vdash n} \pm\ \chi^\lambda(\rho)\ x_1^{\lambda_1+n-1}\ x_2^{\lambda_2+n-2} \cdots x_n^{\lambda_n}$$

where, following the classical style, the notation on the right indicates that the sum is over not just all $\lambda\vdash n$, but also all permutations of the x_i, with the sgn of the permutation included.

Let us compute $\chi^{21}(3)$ - the value of the character of the two-dimensional irreducible representation of S_3 acting on the class consisting of 3-cycles. We already know (p. 93) the answer is -1.

$n=3$, so the variables are x_1, x_2, x_3

$\rho=(3)$ so $s_\rho = s_3 = x_1^3 + x_2^3 + x_3^3$

$\Delta = (x_1-x_2)\ (x_1-x_3)\ (x_2-x_3)$

so the left side is $(x_1^3+x_2^3+x_3^3)(x_1-x_2)(x_1-x_3)(x_2-x_3)$.

The right hand side is, first of all, a sum over all partitions λ of $n=3$.

For $\lambda=(1,1,1)$ we get $(\lambda_1+n-1, \lambda_2+n-2, \lambda_3+n-3) = (3,2,1)$

so this partition contributes to the total the terms

$\chi^{1^3}(3)\ (x_1^3x_2^2x_3+x_2^3x_3^2x_1+x_3^3x_1^2x_2-x_1^3x_3^2x_2-x_2^3x_1^2x_3-x_3^3x_2^2x_1)$.

For $\lambda=(2,1,0)$ we get $(\lambda_1+n-1, \lambda_2+n-2, \lambda_3+n-3) = (4,2,0)$

so this partition contributes to the total the terms

$$\chi^{21}(3)\ (x_1^4x_2^2+x_2^4x_3^2+x_3^4x_1^2-x_1^4x_3^2-x_2^4x_1^2-x_3^4x_2^2) \quad .$$

For $\lambda=(3,0,0)$ we get $(\lambda_1+n-1, \lambda_2+n-2, \lambda_3+n-3) = (5,1,0)$

so this partition contributes to the total the terms

$$\chi^{3}(3)\ (x_1^5x_2+x_2^5x_3+x_3^5x_1-x_1^5x_3-x_2^5x_1-x_3^5x_2) \quad .$$

Hence $\chi^{21}(3)$ appears on the right hand side as the coefficient of $x_1^4x_2^2$. Computation of the left hand side shows that this coefficient is -1.

Obviously one of the problems in computing characters is to find a less messy systematization of this procedure. This systematization, together with the procedure that arises from the final identity, $h_n(xy) = \sum \{\pi\}(x)\{\pi\}(y) = \sum \varphi_\pi(x)\langle\pi\rangle(y)$ is the technique of Young Diagrams, which is the subject of the next section.

4. Methods of Calculation. Young Diagrams.

The object of this chapter is to carry out several calculations, and derive some facts about the representation theory of the symmetric group from them. In particular, we first compute the degrees of the irreducible representations of S_n, and then give, without proof, Frame's result on computing these via "hooks". Then the method of computing general values of the characters $\chi^\lambda(\mu)$ is given. Finally, the transition matrices between the Schur functions and the h_π, a_π, $\langle\pi\rangle$, and f_π are given, and we discuss the phenomenon of the triangularity of these matrices.

The calculation of characters of symmetric groups involves finding a simple way to use the Frobenius Character Formula, derived in the last section:

$$s_\rho(X)\ \Delta(X) = \sum_{\substack{\sigma\in S_n \\ \lambda\vdash n}} \operatorname{sgn}(\sigma)\ \chi^\lambda(\rho)\ X_{\sigma(1)}^{\lambda_1+n-1} \cdots\cdots X_{\sigma(n)}^{\lambda_n}$$

where $\rho\vdash n$, and $\Delta(X) = \prod_{1\le i<j\le n} (X_i - X_j)$.

Consider a monomial $X_{i_1} X_{i_2} \dots X_{i_q}$ with the subscripts chosen from the set $\{1,\dots,n\}$. Such a monomial is called a <u>lattice permutation</u> if in the first i factors, the number of X_1's occuring is $\geq$ the number of X_2's occuring $\geq$ the number of X_3's, etc., for all i. For example, the following are lattice permutations of $X_1^{\ 3} X_2 X_3$:

$$X_1^{\ 3} X_2 X_3 \qquad X_1^{\ 2} X_2 X_1 X_3 \qquad X_1^{\ 2} X_2 X_3 X_1$$

$$X_1 X_2 X_1^{\ 2} X_3 \qquad X_1 X_2 X_3 X_1^{\ 2} \qquad X_1 X_2 X_1 X_3 X_1$$

and the following are not:

$$X_2 X_1^{\ 3} X_3 \qquad X_1^{\ 3} X_3 X_2 \qquad X_1 X_3 X_1 X_2 X_1 \quad .$$

Recall (p.29) that, given a partition $\lambda \vdash n$, $\lambda = (\lambda_1, \lambda_2, \dots, \lambda_n)$, $\lambda_1 \geq \lambda_2 \geq \dots \geq \lambda_n \geq 0$, its Young diagram is a pattern of boxes with λ_1 boxes in the first row, λ_2 in the second, etc.

$\lambda = (5,4,1,1)$

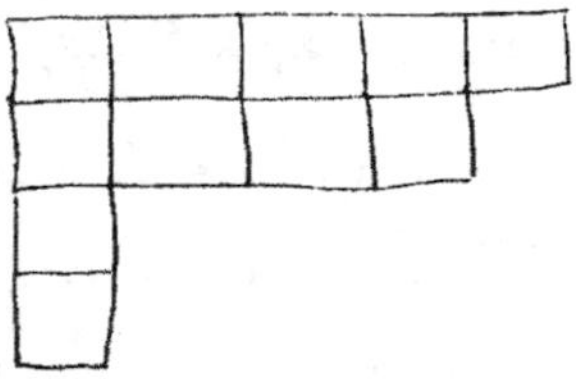

5

4

1

1

A <u>Young tableau</u>, of shape $\lambda = (\lambda_1,\ldots,\lambda_n)$ is obtained by taking the Young diagram λ and filling in the boxes with the numbers $1,2,\ldots,n$ in some order, one number to each box. For example

1	4	11	9	3
2	5	6	10	
7				
8				

and

9	10	6	8	7
3	4	5	11	
2				
1				

A <u>standard</u> Young tableau is one where in each row and each column, the entries are increasing. For example, the two tableaux above are not standard. But the following two are standard:

1	2	3	4	5
6	7	8	9	
10				
11				

1	4	6	8	9
2	5	10	11	
3				
7				

.

<u>Lemma</u>: Given a monomial $\mu = X_1^{\lambda_1}X_2^{\lambda_2}\ldots X_n^{\lambda_n}$, $\lambda_1 \geq \lambda_2 \geq \ldots \geq \lambda_n$, $\sum \lambda_i = n$, the number of lattice permutations of μ is equal to the number of standard tableaux of shape $(\lambda_1,\ldots,\lambda_n)$.

<u>Proof</u>: By example. Suppose $\mu = X_1^5X_2^4X_3$. Given the lattice permutation $X_1^3X_2X_1X_2^2X_3X_1X_2$, label the terms in order

1	2	3	4	5	6	7	8	9	10
X_1	X_1	X_1	X_2	X_1	X_2	X_2	X_3	X_1	X_2

Fill out the tableau (5,4,1) by putting the numbers written above the X_1's in the 1st row, those above the X_2's in the 2nd, etc.

Conversely, given a standard tableau

1	3	5	6	9
2	7	8	10	
4				

write out the monomial with an X_1 in the 1st, 3^d, 5th, 6th and 9th position, X_2 in 2nd, 7th, 8th and 10th, and X_3 in the 4th. This yields $X_1X_2X_1X_3X_1^2X_2^2X_1X_2$.

These two processes are inverse, giving a one-one correspondence between standard tableaux and lattice permutations. ■

<u>Theorem</u>: Let $\lambda \vdash n$. The degree of the irreducible representation χ^λ is the number of standard tableaux of shape λ.

Proof: The degree is $\chi^\lambda(1^n)$. (More accurately, since we only know so far that $\chi^\lambda = \Theta^{-1}(\{\lambda\})$ is $\pm$ an irrep, the degree is the absolute value of $\chi^\lambda(1^n)$. But it will follow from this calculation that $\chi^\lambda(1^n)$ is positive for all λ.)

By the Frobenius Character formula

$$(X_1+\dots+X_n)^n \prod_{i<j} (X_i-X_j) = \sum_{\substack{\sigma\in S_n \\ \lambda\vdash n}} \operatorname{sgn}(\sigma)\, \chi^\lambda(1^n)\, X_{\sigma(1)}^{\lambda_1+n-1} \cdots\cdots X_{\sigma(n)}^{\lambda_n} .$$

The product $\prod_{i<j} (X_i-X_j)$ can be expanded as

$$\Delta(X) = \sum \pm X_1^{n-1} \cdots\cdots X_{n-1}$$

where the sum is over all permutations $\sigma \in S_n$ of the subscripts with the $\mathrm{sgn}(\sigma)=\pm 1$ attached.

We want to calculate $\chi^{\lambda}(1^n)$ which, on the right side, is the coefficient of, in particular,

$$X_1^{\lambda_1+n-1} \cdot X_2^{\lambda_2+n-2} \cdots X_n^{\lambda_n} = (X_1^{\lambda_1} X_2^{\lambda_2} \cdots X_n^{\lambda_n})(X_1^{n-1} X_2^{n-2} \cdots X_{n-1}).$$

So we may ask for the ways this monomial appears on the left in the product

$$(X_1+\cdots+X_n)^n \sum \pm X_1^{n-1} X_2^{n-2} \cdots X_{n-1} .$$

Let us expand this product, starting with $\Delta(X)$ and successively multiplying by $(X_1+\cdots+X_n)$ n times. At each stage, $(X_1+\cdot\cdot+X_n)^i \Delta(X)$, the function obtained is still alternating (i.e., if the X's are permuted by an odd permutation $\sigma \in S_n$, the sign changes by -1). This occurs since $\Delta(X)$ is alternating and the product of a symmetric function by an alternating function is alternating. Hence writing $(X_1+\cdots+X_n)^i \Delta(X)$ as a sum of monomials ((integer coefficient)$X_1^{i_1} \cdots X_n^{i_n}$) in each monomial with nonzero coefficient, the powers of different X's have to be different.

Let $\tau = X_1^{j_1} X_2^{j_2} \cdots X_n^{j_n}$ be one such term in $(X_1+\cdot\cdot+X_n)^i \Delta(X)$ with distince indices $j_1,\ldots,j_n$. Multiply $(X_1+\cdot\cdot+X_n)^i \Delta(X)$ by $(X_1+\cdots+X_n)$ to get to the next step $(X_1+\cdot\cdot+X_n)^{i+1} \Delta(X)$. For τ as above, and any k, look at the resulting term $X_k \tau = X_1^{j_1'} X_2^{j_2'} \cdot\cdot X_n^{j_n'}$.

Each $j_q = j'_q$ except for $q=k$ where $j'_k = j_k + 1$. Then either two of the j' are equal, in which case, since $(X_1 + .. + X_n)^{i+1}\Delta(X)$ is alternating, the coefficient of $X_k \tau$ is zero, or the $j'_1, j'_2, \ldots, j'_n$ still exhibit the same pattern as the $j_1, j_2, \ldots, j_n$ in the sense that if j_k is the largest of $j_1, \ldots, j_n$, then j'_k is the largest of $j'_1, \ldots, j'_n$, and etc. for the second largest, third largest ... In particular, to find out which terms contribute to $X_1^{\lambda_1 + n-1} \ldots X_n^{\lambda_n}$ in $(X_1 + \ldots + X_n)^n \Delta(X)$, one only need start with $(X_1^{n-1} \ldots X_{n-1})$ and ask how this term changes each time one multiplies in a factor of $(X_1 + \ldots + X_n)$ At each step, one is given $(X_1^{n-1} \ldots X_{n-1})(X_1^{\mu_1} X_2^{\mu_2} \ldots)$, $\mu_1 \geq \mu_2 \geq \ldots$ After multiplying by some X_i, one gets $(X_1^{n-1} \ldots X_{n-1})(X_1^{\mu'_1} X_2^{\mu'_2} \ldots)$, $\mu'_1 \geq \mu'_2 \geq \ldots$ Hence in the resulting $\chi^\lambda(1^n)(X_1^{n-1} \ldots X_{n-1})(X_1^{\lambda_1} X_2^{\lambda_2} \ldots)$ the coefficient $\chi^\lambda(1^n)$ is the number of ways of arriving at the term $(X_1^{n-1} \ldots X_{n-1})(X_1^{\lambda_1} X_2^{\lambda_2} ..)$, which is the number of lattice permutations of $X_1^{\lambda_1} X_2^{\lambda_2} \ldots X_n^{\lambda_n}$. (In particular $\chi^\lambda(1^n)$ is positive.) ∎

For example, for n=3, the standard tableaux are:

Shape	Standard Tableaux	Degree of associated character
(3)	$\begin{array}{\|c\|c\|c\|}\hline 1 & 2 & 3 \\ \hline\end{array}$	$\deg \chi^{3} = 1$
(21)	$\begin{array}{\|c\|c\|}\hline 1 & 2 \\ \hline 3 \\ \cline{1-1}\end{array}$	$\deg \chi^{21} = 2$
	$\begin{array}{\|c\|c\|}\hline 1 & 3 \\ \hline 2 \\ \cline{1-1}\end{array}$	
(111)	$\begin{array}{\|c\|}\hline 1 \\ \hline 2 \\ \hline 3 \\ \hline\end{array}$	$\deg \chi^{111} = 1$

For n=4, λ = (2,2), say, deg χ^{22} = 2 since the only two tableaux of this shape which are standard are

$$\begin{array}{|c|c|}\hline 1 & 2 \\ \hline 3 & 4 \\ \hline\end{array} \quad \text{and} \quad \begin{array}{|c|c|}\hline 1 & 3 \\ \hline 2 & 4 \\ \hline\end{array}$$

Given λ n, we have defined the hook number, H_λ, of λ to be this degree χ^λ. This name comes from the following simple method for computing H_λ (due to Frame ([17])). Given a Young diagram, we assign to each square in it a positive integer, the <u>length</u> of the <u>hook subtended</u> by the square. This "hook" consists of all the squares lying either directly to the right, or directly below, the square in question, together with the square itself. The length is the number of squares involved. For example, the square

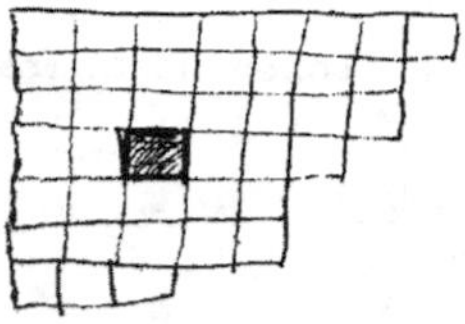

subtends the hook

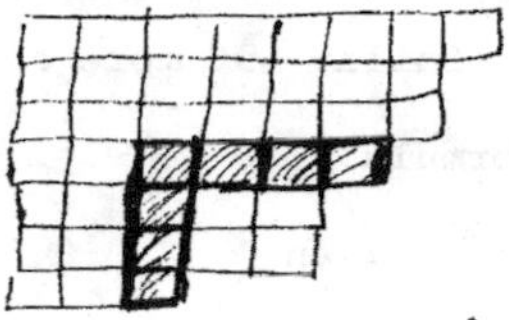

which has length 7. In the following diagrams, the hook numbers h_{ij} are given in the squares.

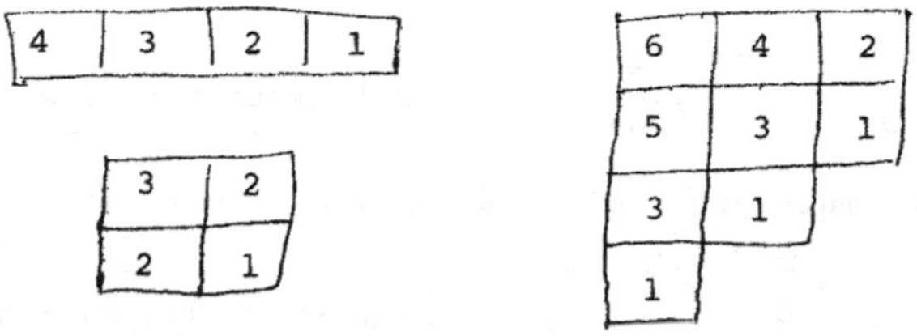

<u>Theorem</u>: $H_\lambda = \dfrac{n!}{\prod_{i,j} h_{ij}}$

<u>Proof</u>: (See [8], [17], or [12]). ∎

A "natural" proof of this has yet to be discovered. $n!$ counts the number of all tableaux of the given shape, and H_λ the number of standard tableaux. It should be that $\prod h_{ij}$ counts the number of some objects and there is an algorithm which to each pair consisting of one object and one standard tableau, assigns a general tableau, and vice-versa. No such algorithm is now known. (Note that the obvious generalization to three dimensional tableaux is false.)

To calculate characters $\chi^\lambda(\rho)$ for general $\lambda, \rho \vdash n$ involves a similar but more complicated use of the Frobenius Character formula.

Given

$$s_\rho\ \Delta(X) = \sum \pm \chi^\lambda(\rho)\ X_1^{\lambda_1+n-1} \dots X_n^{\lambda_n} \quad .$$

Let $\rho = (r_1, r_2, \dots)$ so $s_\rho = s_{r_1} s_{r_2} \dots$, $s_{r_i} = X_1^{r_i} + \dots + X_n^{r_i}$ and

recall $\Delta(X) = \sum \pm X_1^{n-1} X_2^{n-2} \dots X_{n-1}$. In calculating $s_\rho\ \Delta(X)$, we start with $\Delta(X)$ and multiply, one at a time, by the terms s_{r_i}. Suppose we are at stage $\Delta(X) s_{r_1} s_{r_2} \dots s_{r_{i-1}}$, and the multiplication by $s_{r_i} = s_r = X_1^r + \dots + X_n^r$ is next. The partial product is an alternating function, so a sum of monomials in each of which, X's with different subscripts have different exponents. Consider the term

$$(X_1^{n-1} X_2^{n-2} \dots X_{n-1})(X_1^{\alpha_1} X_2^{\alpha_2} \dots X_q^{\alpha_q} X_{q+1}^{\alpha_{q+1}} \dots X_{q+i}^{\alpha_{q+i}} \dots X_n^{\alpha_n})$$

where $\alpha_1 \geq \alpha_2 \geq \dots \geq \alpha_n$. When multiplying by s_r consider the term which arises by multiplying this given term by X_{q+i}^r. If as a result of the multiplication, two different X_j end up with equal exponents, the coefficient of the resulting term will be zero in $\Delta(X) s_{r_1} \dots s_r$ since this is an alternating function.

Hence, for some q,

$$\alpha_q + n - q > \alpha_{q+i} + n - (q+i) + r > \alpha_{q+1} + n - (q+1) \quad .$$

Hence in $\Delta(X)s_{r_1}\ldots s_r$, we get the term

$$(X_1^{n-1}\ldots X_q^{n-q}X_{q+i}^{n-q-1}X_{q+1}^{n-q-2}\ldots X_{n-1}) \cdot$$

$$(X_1^{\alpha_1}\ldots X_q^{\alpha_q}X_{q+i}^{\alpha_{q+i}+r-i+1}X_{q+1}^{\alpha_{q+1}+1}\ldots X_{q+i-1}^{\alpha_{q+i-1}+1}X_{q+i+1}^{\alpha_{q+i+1}}\ldots)$$

There is a corresponding term where the subscripts are in increasing order - with a minus sign in front if i is even. This is the term considered at the next stage, where we will have to keep track of the $\pm$ sign that has arisen here.

In this way, we make the transition from $\Delta(X)s_{r_1}\ldots s_{r_{i-1}}$ to $\Delta(X)s_{r_1}\ldots s_{r_{i-1}}s_r$, and calculate each way a term of type $(X_1^{n-1}\ldots X_{n-1})(X_1^{\alpha_1}\ldots X_n^{\alpha_n})$, $\alpha_1 \geq \alpha_2 \geq \ldots \geq \alpha_n \geq 0$ can arise. At the end, we have all the ways that the term $(X_1^{n-1}\ldots X_{n-1})(X_1^{\lambda_1}\ldots X_n^{\lambda_n})$ can arise - and the coefficient of this term is $\chi^\lambda(\rho)$, which is the number we wished to calculate.

The idea now is to represent the procedure by associating to the term $(X_1^{n-1}\ldots X_{n-1})(X_1^{\alpha_1}\ldots X_n^{\alpha_n})$, $\alpha_1 \geq \ldots \geq \alpha_n \geq 0$, the Young diagram $(\alpha_1,\ldots,\alpha_n)$. $\Delta(X)$ is thus represented by the empty diagram. Passing from $\Delta(X)s_{r_1}\cdots s_{r_{i-1}}$ to $\Delta(X)s_{r_1}\cdots s_{r_{i-1}}s_r$ is interpreted as adding to each of the (signed!) diagrams of $\Delta(X)s_{r_1}\cdots s_{r_{i-1}}$, r squares "in all possible ways" to get a set of (signed) diagrams representing the terms of interest in $\Delta(X)s_{r_1}\cdots s_{r_{i-1}}s_r$.

Specifically: in $\Delta(X)s_{r_1}\cdots s_{r_{i-1}}$, multiplying by X_{q+i}^r, the transition is from a term

$$(X_1^{n-1}\cdots X_{n-1})(X_1^{\alpha_1}X_2^{\alpha_2}\cdots X_q^{\alpha_q}X_{q+1}^{\alpha_{q+1}}X_{q+2}^{\alpha_{q+2}}\cdots X_{q+i-1}^{\alpha_{q+i-1}}X_{q+i}^{\alpha_{q+i}}X_{q+i+1}^{\alpha_{q+i+1}}\cdots X_n^{\alpha_n})$$

to a term

$$(X_1^{n-1}\cdots X_{n-1})(X_1^{\alpha_1}X_2^{\alpha_2}\cdots X_q^{\alpha_q}X_{q+1}^{\alpha_{q+i}+r-i+1}X_{q+2}^{\alpha_{q+1}+1}\cdots X_{q+i-1}^{\alpha_{q+i-2}+1}X_{q+i}^{\alpha_{q+i-1}+1}\cdot$$

$$\cdot X_{q+i+1}^{\alpha_{q+i+1}}\cdots X_n^{\alpha_n}) \quad .$$

So the passage is from the Young diagram

$$(\alpha_1, \alpha_2,\ldots,\alpha_q,\alpha_{q+1},\ldots\ldots,\alpha_{q+i},\alpha_{q+i+1},\ldots\ldots,\alpha_n)$$

to the Young diagram

$$(\alpha_1,\alpha_2,\ldots,\alpha_q,\alpha_{q+i}+r-i+1,\alpha_{q+1}+1,\ldots,\alpha_{q+i-1}+1,\alpha_{q+i+1},\ldots,\alpha_n) \quad .$$

The transition from the old diagram to the new can be described as adding r squares in a certain way: to the q+i row, add enough squares to change it from length α_{q+i} to length $\alpha_{q+i-1}+1$, i.e., fill out the row until it exceeds the previous row by one square:

q+i-1 row

q+i row

Then add enough squares to the q+i-1 row to change its length to $\alpha_{q+i-2}+1$ - i.e., enough to exceed the next higher row by one square. Keep doing this until you run out of squares to add. Suppose this exhaustion occurs in the qth row and the result is a Young diagram (i.e., the exhaustion does not occur at one of the times when a row has one more square than a previous row.) Pictorially, a typical addition is

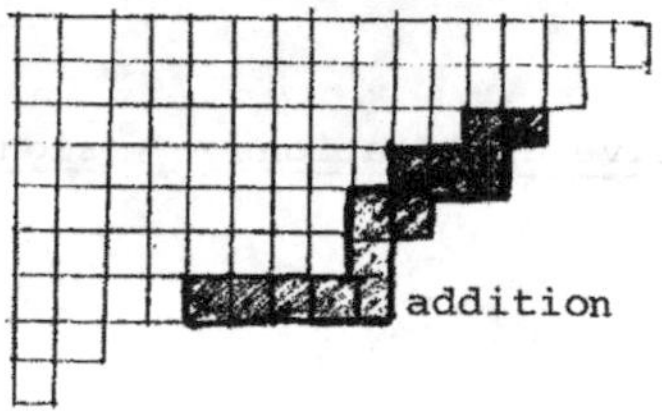

q

q+i

addition

(here r=13 squares are added)

This process is called the <u>regular addition of r squares</u> to the Young diagram $(\alpha_1,\ldots,\alpha_n)$. It is a <u>positive</u> application if the number of rows involved (= i+1) is odd, <u>negative</u> if the number of rows involved is even. (This is to keep track of the $\pm$ signs).

Thus our algorithm for computing $\chi^{\lambda}(\rho)$ can be described:

Let $\lambda = (\lambda_1,\ldots,\lambda_n)$, and $\rho = (r_1,\ldots,r_n)$ be partitions of n. Compute all ways of starting with the empty Young diagram and adding, in a regular fashion, first r_1 squares, then r_2, then $r_3,\ldots$, then r_n so that the result is the Young diagram λ. For each such way, count (-1) if there are an odd number of negative applications, (+1) if there are an even number (or zero) negative applications. Add up the signed total, the result is $\chi^{\lambda}(\rho)$.

For example: $\chi^{331}(2221) = -3$

Here $\lambda=(3,3,1)$ and $\rho=(2,2,2,1)$. The procedure is first to add 2 squares, then 2 more, then 2 more, then 1, to get the pattern (3,3,1). If the first 2 squares are labeled "a", the next 2, "b", the next 2, "c", and the last, "d", the ways of doing this can be indicated as follows:

	negative applications	sign
$\begin{matrix} a & a & c \\ b & b & c \\ d & & \end{matrix}$	one (c)	-1
$\begin{matrix} a & b & b \\ a & c & c \\ d & & \end{matrix}$	one (a)	-1
$\begin{matrix} a & b & c \\ a & b & c \\ d & & \end{matrix}$	three (a,b,c)	-1

total = $-3 = \chi^{331}(2221)$

Another example: $\chi^{(832)}(5422) = 2$

Allowable patterns	Negative applications	Sign
(pattern 1)	two (a,b)	+1
(pattern 2)	zero	+1

Pattern 1:

a	a	a	a	c	c	d
a	b	b				
b	b					

Pattern 2:

a	a	a	b	b	b	b
a	c	c				
a	d					

Total = 2

From this algorithm we can deduce certain general results on characters.

a) $\chi^{n}(\rho) = 1$ for all ρ

Proof: There is exactly one way to regularly add $\rho=(r_1,\ldots,r_n)$ squares to get the diagram

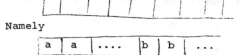

Namely

a	a	....	b	b	...

This involves zero negative applications. ∎

b) $\chi^{1^n}(\rho) = \pm 1$ for all ρ

Proof: There is only one way to regularly add $\rho=(r_1,\ldots,r_n)$ squares to get a vertical column. For each r_i which is even this involves a negative application.

Hence for $\sigma \in S_n$,

$$\chi^{1^n}(\sigma) = \begin{cases} +1 & \text{if } \sigma \text{ involves an even number of cycles of even length} \\ -1 & \text{if } \sigma \text{ involves an odd number of cycles of even length} \end{cases}$$

Hence $\{\sigma \mid \chi^{1^n}(\sigma) = 1\}$ is exactly the alternating group A_n, since a cycle og even length is a product of an odd number permutations and vice-versa. ∎

Of course, we already knew both of these facts. But the next fact is new.

c) $\chi^{\lambda}(\rho) = \chi^{1^n}(\rho)\,\chi^{\lambda'}(\rho)$ (λ' being the conjugate of λ)

<u>Proof</u>: In interchanging the rows and columns of a Young diagram, a regular application of squares remains regular. The number of rows plus the number of columns equals r+1, where r is the number of squares added. Hence if r is odd, and one has a negative application to a diagram, so an even number of rows are involved, the corresponding application to the conjugate diagram is also negative. And a positive application remains positive. If r is even, then a negative application turns into a positive application, and vice-versa.

So for each way of building λ from $\rho=(r_1,\ldots,r_n)$, one can build λ'. If ρ is a positive class, i.e., if ρ has an even number of even cycles, there is no change in sign. If ρ is negative, there is a change in sign. ∎

An alternative way to write this theorem is $a_n * \chi^{\lambda} = \chi^{\lambda'}$, all $\lambda \vdash n$. Thus, explicitly, the map $\theta : \Lambda \longrightarrow \Lambda$, given by $a_{\pi} \rightsquigarrow h_{\pi}$, all $\pi \vdash n$, all n (p. 139) is given by $\chi^{\lambda} \rightsquigarrow \chi^{\lambda'}$, all $\lambda \vdash n$, all n.

The third calculation to be done is to determine the transition matrix between the bases $\{\{\pi\} \mid \pi \vdash n\}$ and $\{h_\pi \mid \pi \vdash n\}$ of Λ_n. We start with the formula (pp.38,43)

$$h_n(xy) = \sum_{\lambda \vdash n} \{\lambda\}(x)\{\lambda\}(y) = \sum_{\pi \vdash n} h_\pi(x)\langle\pi\rangle(y) \quad .$$

Given a partition $\lambda \vdash n$, consider the problem of expressing $\{\lambda\} = \sum_{\pi \vdash n} r_{\lambda\pi} h_\pi$ in Λ_n. Unique integers $r_{\lambda\pi}$ exist making this an identity, since the set $\{h_\pi \mid \pi \vdash n\}$ is a basis of Λ_n. Inserting this into the formula above, we get

$$\sum_{\lambda \vdash n} \sum_{\pi \vdash n} r_{\lambda\pi} h_\pi(X)\{\lambda\}(Y) = \sum_{\pi \vdash n} h_\pi(X)\langle\pi\rangle(Y) \quad .$$

Since the $\{h_\pi\}$ form a basis, the coefficients of h_π, for a particular π must be equal:

$$\sum_{\lambda \vdash n} r_{\lambda\pi}\{\lambda\}(Y) = \langle\pi\rangle(Y) \quad .$$

So

$$\langle\pi\rangle(Y)\ \Delta(Y) = \sum_{\lambda \vdash n} \sum_{\sigma \in S_n} r_{\lambda\pi}\ (\operatorname{sgn} \sigma)\ Y_{\sigma(1)}^{\lambda_1+n-1} \cdots Y_{\sigma(n)}^{\lambda_n} \quad .$$

Fix λ and π and let us compute $r_{\lambda\pi}$. This integer is the coefficient of, among others, the term $Y_1^{\lambda_1+n-1} \cdots\cdot Y_n^{\lambda_n}$. So the problem is to compute the coefficient of this term in

$$\langle\pi\rangle(Y)\Delta(Y) = \left(\sum_{\sigma \in S_n} Y_{\sigma(1)}^{\pi_1} \cdots\cdot Y_{\sigma(n)}^{\pi_n}\right)\left(\sum_{\tau \in S_n} (\operatorname{sgn} \tau) Y_{\tau(1)}^{n-1} \cdots Y_{\tau(n-1)}^{1}\right)$$

I.e., we have to decide for which $\sigma,\tau \in S_n$, the corresponding terms on the right side here multiply to give $Y_1^{\lambda_1+n-1}\ldots.Y_n^{\lambda_n}$. So we start with $Y_1^{\lambda_1+n-1}\ldots.Y_n^{\lambda_n}$ and ask for all ways of decomposing it into a product $(Y_{\sigma(1)}^{\pi_1}\ldots Y_{\sigma(n)}^{\pi_n})\ (Y_{\tau(1)}^{n-1}\ldots Y_{\tau(n-1)}^{1})$. To each such way a $\pm$ sign (= sgn τ) is attached, and $r_{\lambda\pi}$ is the sum of these ± 1.

In terms of Young diagrams, start with the diagram $(\lambda_1,\ldots,\lambda_n)$ representing the term $Y_1^{\lambda_1}\ldots\ldots Y_n^{\lambda_n}$. Add in n-1 squares to the first row, n-2 to the second, etc., to get the diagram of $Y_1^{\lambda_1+n-1}\ldots\ldots Y_n^{\lambda_n}$. Now, dividing by $Y_{\tau(1)}^{n-1}Y_{\tau(2)}^{n-2}\ldots Y_{\tau(n-1)}^{1}$ amounts to subtracting n-1 squares from the $\tau(1)$ row, n-2 squares from the $\tau(2)$ row, etc., where the only constraint on τ is that a row from which k squares are to be subtracted must contain at least k squares.

This expression for $\{\lambda\}$ in terms of h_π's can be expressed formally as follows. Given a partition $\lambda = (\lambda_1,\ldots,\lambda_n)$ of n and a permutation $\sigma \in S_n$, write $\sigma*\lambda$ for the partition $(\lambda_1+n-\sigma(n),\ \lambda_2+(n-1)-\sigma(n-1),\ldots,\lambda_n+1-\sigma(1))$ of n. (If any of these numbers are negative, $\sigma*\lambda$ is undefined.) Then the expression derived above is

$$\{\lambda\} = \sum_{\sigma\in S_n} \operatorname{sgn}(\sigma)\, h_{\sigma*\lambda}$$

This looks like the expression of some determinant, and in fact it is: $\{\lambda\} = \det | h_{\lambda_s - s+t} |$. I.e.,

$$\{\lambda\} = \det \begin{vmatrix} h_{\lambda_1} & h_{\lambda_1+1} & h_{\lambda_1+2} & \cdots & h_{\lambda_1+n} \\ h_{\lambda_2-1} & h_{\lambda_2} & h_{\lambda_2+1} & & h_{\lambda_2+n-1} \\ \cdots\cdots\cdots & & & & \\ h_{\lambda_n-n} & & & & h_{\lambda_n} \end{vmatrix}$$

(where, by definition, $h_0 = 1$, and $h_q = 0$ for $q<0$.) This expression for $\{\lambda\}$ is the Jacobi-Trudi equation.

We carry out the computations for the case n=3.

If $\lambda = (111)$, we start with  , add in (2,1,0) squares

to get 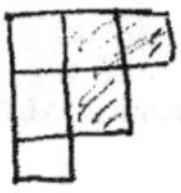 and then there are four ways to subtract two

squares from some row and one from another:

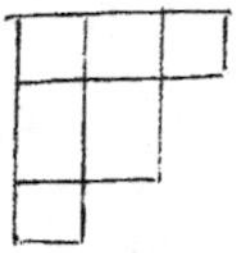 minus $\begin{matrix}2\\1\\0\end{matrix}$ yielding 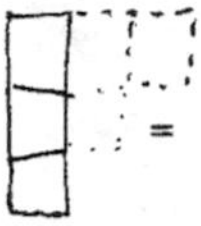 = (1,1,1) $\mathrm{sgn}\binom{210}{210}=+1$

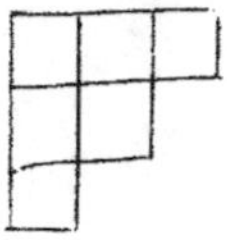 minus $\begin{matrix}2\\0\\1\end{matrix}$ yielding  = (2,1,0) $\mathrm{sgn}\binom{210}{201}=-1$

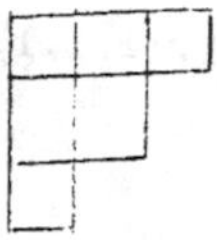 minus $\begin{matrix}1\\2\\0\end{matrix}$ yielding 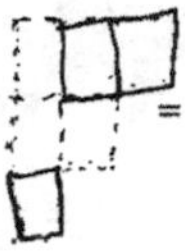 = (2,1,0) $\mathrm{sgn}\binom{210}{120}=-1$

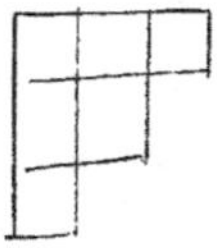 minus $\begin{matrix}0\\2\\1\end{matrix}$ yielding 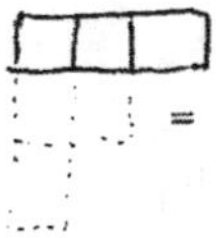 = (3,0,0) $\mathrm{sgn}\binom{210}{021}=+1$

Hence $a_{(111),(111)} = 1$, $a_{(111),(210)} = -2$, $a_{(111),(300)} = 1$.

Similarly, starting with $\lambda = (2,1,0) =$ 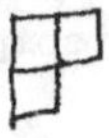 , adding in

(2,1,0) squares gives 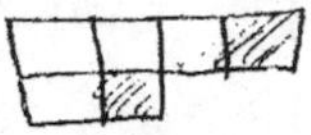 , and there are two ways

to subtract two squares from some row and one from some other:

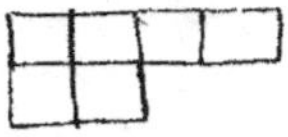 minus $\begin{matrix}2\\1\\0\end{matrix}$ yielding 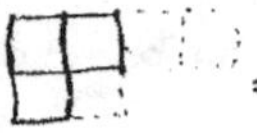 $= (2,1,0) \quad \mathrm{sgn}\binom{210}{210} = +1$

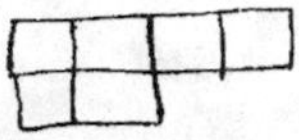 minus $\begin{matrix}1\\2\\0\end{matrix}$ yielding 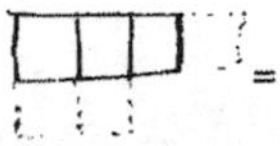 $= (3,0,0) \quad \mathrm{sgn}\binom{210}{120} = -1$

so $a_{(210),(300)} = -1,\ a_{(2,1,0),(2,1,0)} = 1,\ a_{(2,1,0),(3,0,0)} = 0.$

Similarly $a_{(3,0,0),(3,0,0)} = 1,\ a_{(3,0,0),(2,1,0)} = 0, a_{(3,0,0),(1,1,1)} = 0.$

Hence

$$\{3\} = 1h_3 + 0h_{21} + 0h_{111}$$

$$\{21\} = -1h_3 + 1h_{21} + 0h_{111}$$

$$\{111\} = 1h_3 - 2h_{21} + 1h_{111}$$

and so the transition matrix is

$$\begin{pmatrix} 1 & 0 & 0 \\ -1 & 1 & 0 \\ 1 & -2 & 1 \end{pmatrix} .$$

Note in the calculation above, the final transition matrix in triangular! It is a general fact that the transition matrices relating the basis of Schur functions to any of the bases $h_\pi, a_\pi, \langle\pi\rangle$, and f_π are triangular and this follows from some observations on the algorithm above, as we now show.

Definition: On the set $\Pi(n)$ of partitions of an integer n, the natural partial ordering is defined as the least ordering in which, given $\pi_1,\pi_2 \in \Pi(n)$, and $\sigma \in S_n$, with $\pi_1 * \sigma = \pi_2$, then $\pi_1 \leq \pi_2$.

We could also consider the two linear orderings defined by, for $(\lambda_1,\ldots,\lambda_n),(\lambda_1',\ldots,\lambda_n') \in \Pi(n)$,

$$(\lambda_1,\ldots,\lambda_n) >_1 (\lambda_1',\ldots,\lambda_n') \text{ if } \lambda_1=\lambda_1',\ldots,\lambda_i=\lambda_i', \text{but } \lambda_{i+1}>\lambda_{i+1}'$$

and

$$(\lambda_1,\ldots,\lambda_n) >_2 (\lambda_1',\ldots,\lambda_n) \text{ if } \lambda_n=\lambda_n',\ldots,\lambda_{i+1}=\lambda_{i+1}', \text{but } \lambda_i<\lambda_i' .$$

Thus the first is lexicographic by largest part first, and the second is lexicographic by smallest part last (where, note carefully, all partitions of n are taken to have n parts $\lambda_1 \geq \lambda_2 \geq \cdots \geq \lambda_n \geq 0$, and in general the smallest parts λ_n, λ_{n-1}, etc. will be zero).

It is straightforward to show that if $\pi_1 * \sigma = \pi_2$, for $\pi_1,\pi_2 \in \Pi(n)$, and $\sigma \in S_n$, then both $\pi_1 \leq_1 \pi_2$ and $\pi_1 \leq_2 \pi_2$. In particular this shows that the binary relation on π_1,π_2: $\pi_1\sigma=\pi_2$ for some σ, extends to a partial ordering in which $\pi_1 \leq \pi_2$ and $\pi_2 \leq \pi_1$ implies $\pi_1=\pi_2$.

In the algorithm above, $r_{\lambda\pi} = 0$ unless $\lambda \leq \pi$. Thus $\{\lambda\} = \sum_{\pi \geq \lambda} r_{\lambda\pi} h_\pi$, and $\{n\} = h_n$, $\{n-1,1\} = -h_n + h_{n-1,1}$, etc. Thus if the bases $\{\{\pi\} \setminus \pi \vdash n\}$ and $\{h_\pi \setminus \pi \vdash n\}$ are each ordered by any linear ordering which contains the natural order, the transition matrix is triangular. In fact since $r_{\lambda\lambda}=1$, all λ, (clear from the algorithm) the matrix is of the form

(so-called "unipotent").

Proposition: The natural ordering is equivalent to each of the following partial orders on $\Pi(n)$.

$\leq_Y$: Let $1 \leq i \leq j \leq n$, and $\lambda \in \Pi(n)$ be a partition: $\lambda = (\lambda_1, \lambda_2, \ldots, \lambda_n)$. Suppose $\lambda_{i-1} > \lambda_i$ and $\lambda_j > \lambda_{j+1}$. The Young Raising Operator R_{ji} associates to λ the partition $R_{ji}(\lambda) = (\lambda_1, \lambda_2, \ldots, \lambda_i + 1, \lambda_{i+1}, \ldots, \lambda_j - 1, \lambda_{j+1}, \ldots \lambda_n)$. That is, one unit is removed from a small part, the jth, and added to a larger part, the ith. We say, for two partitions λ, λ' of n, that $\lambda \leq_Y \lambda'$ if λ' is obtained from λ by applying a finite series of Young Raising Operators.

$\leq_D$: Consider the class of convex functions from the set of integers $\{1, \ldots, n\}$ to itself. (<u>Convex</u> means that $f(i+1)+f(i-1)$ is less than or equal to $2f(i)$, for each $i = 2, \ldots, n-1$.) Given a partition $\lambda = (\lambda_1, \ldots, \lambda_n)$ of n, consider the convex function

f_λ given by $f_\lambda(i) = \lambda_1 + \lambda_2 + \ldots + \lambda_i$. Conversely, a convex function f from $\{1,\ldots,n\}$ to itself satisfying $f(n)=n$ is of the form f_λ for $\lambda = (f(1), f(2)-f(1),\ldots,n-f(n-1))$. So the two sets are in one-one correspondence and it makes sense to define for two partitions λ,λ' of n that $\lambda \leq_D \lambda'$ if $f_\lambda \leq f_{\lambda'}$ as functions, i.e., $f_\lambda(q) \leq f_{\lambda'}(q)$ for all $q=1,,,n$.

Proof: It is easy to see that $\leq_Y$ and $\leq_D$ are equivalent, likewise that $\pi \leq_Y \pi'$ implies $\pi \leq \pi'$. What remains essentially is to show that if $\pi * \sigma = \pi'$, then π' is obtained from π by applying a series of raising operators. Let $\pi = (\pi_1,\ldots,\pi_n)$. Add in the numbers $n,n-1,..$ to get $(\pi_1+n,\pi_2+n-1,..,\pi_n)$. The numbers are now all distinct. To get $\pi*\sigma$, one now subtracts $\sigma(n)$ from the first number, $\sigma(n-1)$ from the second, etc. Let $i\in\{1,..,n\}$. Consider the number from which the number i is to be subtracted. Suppose it is larger than the number from which i+1 is to be subtracted. Let p,q, respectively, be the indices of these rows. Let σ' be σ followed by the transposition $(i,i+1)$. Then $\pi*\sigma = R_{qp}(\pi*\sigma')$. Proceeding in this fashion, we can assume $\pi*\sigma$ is obtained by applying a series of raising operators to a partition of the form $\pi*\sigma_1$, where, in computing $\pi*\sigma_1$, the number from which the number i is to be subtracted is less than the number from which i+1 is to be subtracted, for all i. But this can only happen if σ_1 is the identity permutation, so $\pi*\sigma_1 = \pi$. ∎

Another property of the natural ordering is that for $\pi,\lambda \in \Pi(n)$, $\pi \leq \lambda$ if and only if, taking conjugate partitions, $\lambda' \leq \pi'$.

One might conjecture that the natural ordering is also obtained as the intersection of the two linear orderings $\leq_1, \leq_2$ above. But this is false, as (7,4,4,1) is greater than (6,6,2,2) in each of these, but the two are incomparable under the natural order.

This partial ordering on partitions has been investigated by, among others, Doubilet ([14]) who has proved a Mobius Inversion Theorem on it and also by Brylawski ([9]) and Liebler and Vitale ([25]).

Young originally gave the algorithm in terms of his Raising operator, and conversely the expression of the h_π's in terms of the $\{\pi\}$'s by means of a lowering operator. (See [39]).

Now that we have the transition matrices $R = (r_{\lambda\pi})$, $\{\lambda\} = \sum_\pi r_{\lambda\pi} h_\pi$, it is a simple matter to compute the matrix $T = (t_{\lambda\pi})$, $\langle\lambda\rangle = \sum_\pi t_{\lambda\pi}\{\pi\}$. To find $t_{\lambda\pi}$, dot this equation with $\{\mu\}$:

$$(\langle\lambda\rangle, \{\mu\}) = \sum_\pi t_{\lambda\pi}(\{\pi\},\{\mu\}) = t_{\lambda\mu} \qquad \text{(using the orthonormality of the } \{\mu\}\text{'s)}$$

But $\{\mu\} = \sum_{\pi} r_{\mu\pi} h_{\pi}$ so $(\langle\lambda\rangle, \{\mu\}) = \sum_{\pi} r_{\mu\pi} (\langle\lambda\rangle, h_{\pi}) = r_{\mu\lambda}$,

since the h_{π}'s and $\langle\lambda\rangle$'s are dual bases. Hence $t_{\lambda\mu} = r_{\mu\lambda}$, so the matrix T is the transpose of R, a fact traditionally known as Kostka's Theorem. In particular T is also triangular.

Now the transition $\langle\lambda\rangle = \sum_{\mu} u_{\lambda\mu} h_{\mu}$ is just the product of the transitions $\langle\lambda\rangle = \sum t_{\lambda\pi}\{\pi\}$, and $\{\pi\} = \sum r_{\pi\mu} h_{\mu}$ so the matrix $U = (u_{\lambda\pi}) = T \cdot R = R^{\text{transpose}} \cdot R$. It is a general fact (and easy to prove) that the product of a matrix by its transpose is a symmetric matrix. Hence in the expression $\langle\lambda\rangle = \sum u_{\lambda\mu} h_{\mu}$, we have $u_{\lambda\mu} = u_{\mu\lambda}$ for all λ, μ.

Next is the transition matrix $W = (w_{\lambda\pi})$, $\{\lambda\} = \sum w_{\lambda\pi} a_{\pi}$. Recall $a_n * \{\lambda\} = \{\lambda'\}$, the conjugate partition, and $a_n * h_{\pi} = a_{\pi}$ and $a_n * a_{\pi} = h_{\pi}$. Thus $a_n * \{\lambda\} = \sum w_{\lambda\pi} a_n * a_{\pi}$, so $\{\lambda'\} = \sum w_{\lambda\pi} h_{\pi}$. Hence $w_{\lambda\pi} = r_{\lambda'\pi}$. Similarly $f_{\pi} = a_n * \langle\pi\rangle$, so this same sort of calculation gives the Schur functions in terms of the f_{π}.

Recall that passing from a partition to its conjugate just turns the partial ordering upside down. Hence the fact that $(r_{\lambda\pi})$ is triangular implies that $(w_{\lambda\pi})$ is also.

Notice in particular the following phenomenon: for each $\lambda \in \Pi(n)$, $(a_{\lambda'}, h_\lambda) = 1$. Proof: inverting the matrix R one still gets a triangular matrix with diagonal entries 1, thus $h_\lambda = \{\lambda\} + \sum_{\pi > \lambda} r_{\lambda\pi}\{\pi\}$. Taking the inner product with a_n, this becomes $a_\lambda = \{\lambda'\} + \sum_{\pi>\lambda} r_{\lambda\pi}\{\pi'\}$. Relabel λ' by λ, π' by π, and the result is $a_{\lambda'} = \{\lambda\} + \sum_{\lambda>\pi} r_{\lambda'\pi}\{\pi\}$. Hence $(h_\lambda, a_{\lambda'}) = (\{\lambda\},\{\lambda\}) = 1$. QED.

This fact again gives the natural correspondence between partitions λ of n and irreps $\{\lambda\}$ of S_n. Indeed, a standard proof (e.g., in [12] [39]) that the correspondence exists is first to compute $(h_\lambda, a_{\lambda'})=1$, (by using the Mackey Theorem, say) and then to observe that since h_λ and $a_{\lambda'}$ are both actual representations, this fact implies that they have one and only one irrep in common. This irrep is then labeled χ^λ. The main problem [redacted] of course, is to show that every irrep of S_n arises in this way for a unique λ.

In the approach of Alfred Young (for which see [39]), the fact that $(h_\lambda, a_{\lambda'})=1$ comes up in his construction of idempotents in $\mathbb{C}[S_n]$ - the Young symmetrizers. For each Young tableau of shape $\lambda \vdash n$, there is constructed an element $P(\lambda) \in \mathbb{C}[S_n]$ - the "positive symmetric group on the rows of λ" and an element

$N(\lambda) \in \mathbb{C}[S_n]$ - the "negative symmetric group on the columns of λ" and then one shows that $\{ \sum_{\substack{\text{all Young} \\ \text{tableaux of} \\ \text{shape } \lambda}} P(\lambda)N(\lambda) \mid \lambda \vdash n\}$

is an orthogonal set in $\mathbb{C}[S_n]$.

The combinatorial aspect of the representation theory of the symmetric group is the study of these transition matrices, $(r_{\lambda\pi})$, $(t_{\lambda\pi})$, etc. An entirely different approach to the subject (in, e.g., Stanley [43], and Rota [38]) starts with symmetric functions and the symmetry of the matrix $W = (w_{\pi\lambda})$, $\langle\pi\rangle = \sum w_{\pi\lambda} h_\lambda$ and constructs bodily the matrix R for which $W = R^{\text{transpose}} \cdot R$. This is done by the ingenious Knuth Algorithm. Then R is used to define Schur functions and their properties are derived. In this approach, the whole notion of group representation can be completely ignored, thus shortening the exposition. However, certain notions which now appear natural to us, may look ad hoc in that context. On the other hand, in that approach, one has very natural interpretations of the matrix elements $r_{\lambda\pi}$, $w_{\lambda\pi}$, etc., which we could derive, but might not have occurred to us, had not the combinatorial approach first found them. It seems the final synthesis of the two approaches has yet to be written.

BIBLIOGRAPHY

[1] Adams, J.F., λ-Rings and Ψ-Operations, (unpublished lecture, 1961)

[2] Adams, J.F., Lectures on Lie Groups, 1969, Benjamin

[3] Atiyah, M., Power Operations in K-Theory, Quart. J. Math., (2) 17 (1966), 165-93. (Also reprinted in Atiyah, M., K-Theory, 1967, Benjamin

[4] Atiyah, M., and D.O.Tall, Group Representations, -Rings, and the J-homomorphism Topology, 8, 1969, 253-97

[5] Bergman, G.M., Ring Schemes: The Witt Scheme, Chapter 26 in D. Mumford Lectures on Curves on an Algebraic Surface, Princeton, 1966

[6] Berthelot, P., Generalities sur les λ-Anneaux. Expose V in the Seminaire de Geometrie Algebrique, Springer-Verlag Lecture Notes in Mathematics 225, 1972

[7] Birkhoff, G., and S. MacLane, A Survey of Modern Algebra, 3d Edition, 1965, MacMillan Co.

[8] Boerner, H., Representations of Groups, 1970, North-Holland

[9] Brylawski, T., The Lattice of Integer Partitions, University of North Carolina Dept. of Math. report, 1972

[10] Burroughs, J., Operations in Grothendieck Rings and Group Representations, Math.Dept. Preprint 228, State University of New York at Albany

[11] Cartier, P., Groupes formels associes aux anneaux de Witt generalises, C.R. Acad. Sc. Paris, t.265, 1967, A-49-52

[12] Coleman, A.J., Induced Representations with Applications to S_n and Gl(n), Queen's Papers in Pure and Applied Math No.4, Queen's University, Kingston, Ontario, 1966

[13] Doubilet, P., Symmetric Functions through the Theory of Distribution and Occupancy, (No.VII of G.C. Rota's On the Foundations of Combinatorial Theory) (to appear)

[14] Doubilet, P., An Inversion Formula involving Partitions, (mimeographed notes, 1972)

[15] Dress, A., Representations of Finite Groups, Part 1, The Burnside Ring, (mimeographed notes, Bielefeld, 1971)

[16] Foulkes, H.O., On Redfield's Group Reduction Functions, Canadian J. Math, 15, 1963, 272-84

[17] Frame, J.S., G.de B.Robinson, and R.M.Thrall, The Hook Lengths of $\mathfrak{S}_n$, Canadian J. Math., 6, 1954, 316-325

[18] Grothendieck, A., La Theorie des Classes de Chern, Bull. Soc.Math.France, 86, 1958, 137-54

[19] Grothendieck, A., Classes de Faisceaux et Theoreme de Riemann-Roch, (0, Appendix, in Seminaire de Geometrie Algebrique, Springer-Verlag Lecture Notes in Mathematics No. 225, 1972)

[20] Hall, P., The Algebra of Partitions, Proc. 4th Canad.Math. Congress, Banff, 1957 (1959) 147-159

[21] Harary, F., and E. Palmer, The Enumeration Methods of Redfield, Am.Journal Math., 89, 1967, 373-384

[22] Hawkins, T., The Origins of the Theory of Group Characters, Archive for the History of the Exact Sciences, VII, 2, 1971, 142-70

[23] Koerber, A., Representations of Permutation Groups I, 1971, Springer-Verlag Lecture Notes in Mathematics, No. 240

[24] Lang, S., Algebra, 1965, Addison-Wesley

[25] Liebler, R.A. and M.R.Vitale, Ordering the Partition Characters of the Symmetric Group (to appear)

[26] Littlewood, D., A University Algebra, Heinemann, 1950

[27] Littlewood, D., Plethysm and the Inner Product of S-Functions, J.London Math. Soc., 32, 1957, 18-22

[28] Littlewood, D., The Inner Plethysm of S-Functions, Canadian J. Math., 10, 1958, 1-16

[29] Littlewood, D., The Theory of Group Characters, 2nd Ed., 1958, Oxford

[30] MacMahon, P.A., Combinatory Analysis, Vol. I,II, Cambridge, 1915, Reprinted Chelsea, 1960

[31] MacMahon, P.A., table of the number of partitions of n, n ≤ 200, in G.H.Hardy and S.Ramanujan, Asymptotic Formulae in Combinatory Analysis, Proc.London Math.Soc., 17, 1918, p.114

[32] Milnor, J., Lectures on Characteristic Classes, (mimeographed notes, Princeton, 1957)

[33] Polya, G., Kombinatorische Anzahlbestimmungen fur Gruppen, Graphen, und Chemische Verbindungen, Acta Math, 68, 1937, 145-254

[34] Read, R.C., The Use of S-Functions in Combinatorial Analysis, Canadian J. Math, 20, 1968, 808-841

[35] Redfield, J.H., The Theory of Group Reduced Distributions, Am. J. Math., 49, 1928, 433-55

[36] Robinson, G. de B., Representation Theory of the Symmetric Group, 1961, University of Toronto Press

[37] Rota, G-C., Combinatorial Theory, Notes by L. Guibas, Bowdoin Summer Seminar in Combinatorial Theory, 1971

[38] Rota, G-C., The Combinatorics of the Symmetric Group, Notes from a Conference at George Washington University, June 1972, (to appear)

[39] Rutherford, D.E., Substitutional Analysis, 1948, Edinburgh University Press

[40] Salmon, G., Modern Higher Algebra, 1885, Cambridge

[41] Serre, J.-P., Representations Lineares des Groupes Finis, 1967, Hermann

[42] Snapper, E., Group Characters and Nonnegative Integral Matrices, J. Algebra, 19, 520-35, 1971

[43] Stanley, R.P., Theory and Application of Plane Partitions, Studies in Applied Math., 50, 1971, 167-88

[44] Swinnerton-Dyer, H.P.F., Applications of Algebraic Geometry to Number Theory, Proc. Symp. Pure Math, A.M.S., 1969, Number Theory Institute

[45] Weyl, H., The Classical Groups, 1949, Princeton

[46] van der Waerden, B.L., Modern Algebra, Ungar, 1950

[47] Young, A., Quantitive Substitutional Analysis, I - VII, (published at various times 1901-1934 in Proc. London Math. Soc. - see Rutherford [39])

INDEX OF NOTATION

a_n	2
$K(F)$	6
$\wedge^n V$	6
$\lambda_t(a)$	8
$P_n(s_1,\ldots,s_n;\sigma_1,\ldots,\sigma_n)$	12
$P_{nd}(s_1,\ldots,s_{nd})$	12
$1 + A[[t]]^+$	15
$\wedge$	25
a_π	28
$\wedge_n$	28
π' (conjugate of π)	29
$\sigma \vdash n$	29
$\Pi(n)$	29
h_π	30
$\langle\pi\rangle$	32
s_π	35
$\Delta(X)$	39
$\{\lambda\}$	43
Ψ^n	47
R^ω	50
W_R	55
S_n	59
$R(G)$	70
$CF(G)$	81
K_i	89
L_i	90
$[\pi]$	96
$K\backslash G/H$	99
$B(G)$	107
$SCF(G)$	110
$\mathbb{C}[G]$	115
$R(S)$	128
Θ	130
$s_{\pi_1} * s_{\pi_2}$	137
$s_{\pi_1} \cdot s_{\pi_2}$	137
φ_π	137
θ	139
$\cap_\sigma$, $\cup^\sigma$	145
$\lvert a_{ij} \rvert^\chi$	147
H_π	151
χ^π	151

$\binom{m}{\pi}$	153	$T = (t_{\lambda\pi})$	190
f_π	162	$U = (u_{\lambda\pi})$	191
$R = (r_{\lambda\pi})$	182	$W = (w_{\lambda\pi})$	191
$\sigma^*\lambda$	184	$P(\lambda), N(\lambda)$	192
R_{ij}	189	$\|T\|$ the number of elements in a set T	passim

INDEX

Adams Operators 47

algebraic geometry 52

binomial coefficient, generalized 153

binomial type 9

Brauer's Theorem 101

Burnside Ring 107

Cauchy's Lemma 40

central function 81

centralizer 105

character 84

character ring 84

character table 90

character of product of two groups G H 97

characters of S_n, integrality 134

characters, computation 101

conjectures 100, 113, 135

conjugate partition 29

cycle 124

cycle index 146

cycle structure 124

dot product in $R(S)=\Lambda$ 138

Ferrar's graph 29

finite degree (of an element in a λ-ring) 8

forgotten symmetric function 162

Frobenius Character Formula 163

Frobenius Reciprocity 74

Fundamental Theorem, Rep. Theory of S_n 135

Fundamental Theorem, symmetric functions 2

G-module 61

- - -, map of 61

- - -, isomorphism of 62

G-set, G-map 104

- - -, sum of 106

- - -, product of 106

- - -, symmetric power 106

group algebra 115

group determinant 122

group reduction formula 146

Homogeneous power sum 30

hook number 172

immanent 147

indecomposible 76

induced character formula 96

inner automorphism 105

inner product in R(S) (=∧) 138

irreducible character 84

irrep (=irreducible representation)

isobaric 28

isotypical 78

isotypical component 79

Jacobi-Trudi Equation 184

K-Theory 27

Knuth Algorithm 193

Kostka's Theorem 191

λ-ring 5

- - -, binomial 9

- - -, category of 20

- - -,of central functions 54

- - -, definition 13

- - -, finitary 8

- - -, free on one generater 24

- - -, map of 20

λ-ring, natural operation on 25

- - -, product of two 21

- - -, special 15

- - -, tensor product of two 21

lattice permutation 167

Mackey's Theorem 99

Maschke's Theorem 76

monomial group 103

natural ordering on partitions 187

Newton's Formulas 35

normal dot product 156

normalizer 105

orbit 104

orthogonality relations 91

outer product 127

partition 29

partition, natural ordering on set of 187

permutation matrix 109

plethysm, inner and outer 135

power sums 35

pre-λ-ring 7

pre-Ψ-ring 49

Ψ-ring 49

regular addition of squares 177

representation, linear 60

- - -, conjugate of 69

- - -, decomposible 76

- - -, degree of 60

- - -, dual of 68

- - -, exterior power 68

- - -, faithful 65

- - -, induced 73

- - -, inner product 72

- - -, irreducible 76

- - -, permutation 64

- - -, product of 68

- - -, reducible 76

- - -, regular 64

- - -, sum of 68

- - -, trivial 64

representation of S_n, canonical 64

- - -, alternating 67

representation ring 70

S-functions 44

Schurfunctions 43

Schur's Lemma 77

semidirect product 98

semi-simple 116

simple G-set 105

Splitting Principle 18

standard Young tableau 168

super central function 110

super character 110

super character table 113

symmetric function 2

- - -, elementary 28

- - -, forgotten 162

- - -, Fund. Theorem 2

- - -, hom. power sum 30

- - -, monomial 32

- - -, power sum 35

- - -, Schur 43

symmetric power 46

torsion free ring 49

trace 83

transitive 105

triangularity of transition matrices 189

Verification Principle 27

Waring formula 35

Wedderburn's Theorem 116

Witt vectors 56

wreath product 98

Young diagram 29

Young Raising Operator 189

Young Symmetrizing Operator 119

Young tableau 168

zeta-function 53

Lecture Notes in Mathematics

Comprehensive leaflet on request

Vol. 146: A. B. Altman and S. Kleiman, Introduction to Grothendieck Duality Theory. II, 192 pages. 1970. DM 18,–

Vol. 147: D. E. Dobbs, Cech Cohomological Dimensions for Commutative Rings. VI, 176 pages. 1970. DM 16,–

Vol. 148: R. Azencott, Espaces de Poisson des Groupes Localement Compacts. IX, 141 pages. 1970. DM 16,–

Vol. 149: R. G. Swan and E. G. Evans, K-Theory of Finite Groups and Orders. IV, 237 pages. 1970. DM 20,–

Vol. 150: Heyer, Dualität lokalkompakter Gruppen. XIII, 372 Seiten. 1970. DM 20,–

Vol. 151: M. Demazure et A. Grothendieck, Schémas en Groupes I. (SGA 3). XV, 562 pages. 1970. DM 24,–

Vol. 152: M. Demazure et A. Grothendieck, Schémas en Groupes II. (SGA 3). IX, 654 pages. 1970. DM 24,–

Vol. 153: M. Demazure et A. Grothendieck, Schémas en Groupes III. (SGA 3). VIII, 529 pages. 1970. DM 24,–

Vol. 154: A. Lascoux et M. Berger, Variétés Kähleriennes Compactes. VII, 83 pages. 1970. DM 16,–

Vol. 155: Several Complex Variables I, Maryland 1970. Edited by J. Horváth. IV, 214 pages. 1970. DM 18,–

Vol. 156: R. Hartshorne, Ample Subvarieties of Algebraic Varieties. XIV, 256 pages. 1970. DM 20,–

Vol. 157: T. tom Dieck, K. H. Kamps und D. Puppe, Homotopietheorie. VI, 265 Seiten. 1970. DM 20,–

Vol. 158: T. G. Ostrom, Finite Translation Planes. IV. 112 pages. 1970. DM 16,–

Vol. 159: R. Ansorge und R. Hass. Konvergenz von Differenzenverfahren für lineare und nichtlineare Anfangswertaufgaben. VIII, 145 Seiten. 1970. DM 16,–

Vol. 160: L. Sucheston, Constributions to Ergodic Theory and Probability. VII, 277 pages. 1970. DM 20,–

Vol. 161: J. Stasheff, H-Spaces from a Homotopy Point of View. VI, 95 pages. 1970. DM 16,–

Vol. 162: Harish-Chandra and van Dijk, Harmonic Analysis on Reductive p-adic Groups. IV, 125 pages. 1970. DM 16,–

Vol. 163: P. Deligne, Equations Differentielles à Points Singuliers Reguliers. III, 133 pages. 1970. DM 16,–

Vol. 164: J. P. Ferrier, Seminaire sur les Algebres Complètes. II, 69 pages. 1970. DM 16,–

Vol. 165: J. M. Cohen, Stable Homotopy. V, 194 pages. 1970. DM 16,–

Vol. 166: A. J. Silberger, PGL_2 over the p-adics: its Representations, Spherical Functions, and Fourier Analysis. VII, 202 pages. 1970. DM 18,–

Vol. 167: Lavrentiev, Romanov and Vasiliev, Multidimensional Inverse Problems for Differential Equations. V, 59 pages. 1970. DM 16,–

Vol. 168: F. P. Peterson, The Steenrod Algebra and its Applications: A conference to Celebrate N. E. Steenrod's Sixtieth Birthday. VII, 317 pages. 1970. DM 22,–

Vol. 169: M. Raynaud, Anneaux Locaux Henséliens. V, 129 pages. 1970. DM 16,–

Vol. 170: Lectures in Modern Analysis and Applications III. Edited by C. T. Taam. VI, 213 pages. 1970. DM 18,–

Vol. 171: Set-Valued Mappings, Selections and Topological Properties of 2^X. Edited by W. M. Fleischman. X, 110 pages. 1970. DM 16,–

Vol. 172: Y.-T. Siu and G. Trautmann, Gap-Sheaves and Extension of Coherent Analytic Subsheaves. V, 172 pages. 1971. DM 16,–

Vol. 173: J. N. Mordeson and B. Vinograde, Structure of Arbitrary Purely Inseparable Extension Fields. IV, 138 pages. 1970. DM 16,–

Vol. 174: B. Iversen, Linear Determinants with Applications to the Picard Scheme of a Family of Algebraic Curves. VI, 69 pages. 1970. DM 16,–

Vol. 175: M. Brelot, On Topologies and Boundaries in Potential Theory. VI, 176 pages. 1971. DM 18,–

Vol. 176: H. Popp, Fundamentalgruppen algebraischer Mannigfaltigkeiten. IV, 154 Seiten. 1970. DM 16,–

Vol. 177: J. Lambek, Torsion Theories, Additive Semantics and Rings of Quotients. VI, 94 pages. 1971. DM 16,–

Vol. 178: Th. Bröcker und T. tom Dieck, Kobordismentheorie. XVI, 191 Seiten. 1970. DM 18,–

Vol. 179: Seminaire Bourbaki – vol. 1968/69. Exposés 347-363. IV. 295 pages. 1971. DM 22,–

Vol. 180: Séminaire Bourbaki – vol. 1969/70. Exposés 364-381. IV, 310 pages. 1971. DM 22,–

Vol. 181: F. DeMeyer and E. Ingraham, Separable Algebras over Commutative Rings. V, 157 pages. 1971. DM 16.–

Vol. 182: L. D. Baumert. Cyclic Difference Sets. VI, 166 pages. 1971. DM 16,–

Vol. 183: Analytic Theory of Differential Equations. Edited by P. F. Hsieh and A. W. J. Stoddart. VI, 225 pages. 1971. DM 20,–

Vol. 184: Symposium on Several Complex Variables, Park City, Utah, 1970. Edited by R. M. Brooks. V, 234 pages. 1971. DM 20,–

Vol. 185: Several Complex Variables II, Maryland 1970. Edited by J. Horváth. III, 287 pages. 1971. DM 24,–

Vol. 186: Recent Trends in Graph Theory. Edited by M. Capobianco/ J. B. Frechen/M. Krolik. VI, 219 pages. 1971. DM 18.–

Vol. 187: H. S. Shapiro, Topics in Approximation Theory. VIII, 275 pages. 1971. DM 22,–

Vol. 188: Symposium on Semantics of Algorithmic Languages. Edited by E. Engeler. VI, 372 pages. 1971. DM 26,–

Vol. 189: A. Weil, Dirichlet Series and Automorphic Forms. V. 164 pages. 1971. DM 16,–

Vol. 190: Martingales. A Report on a Meeting at Oberwolfach, May 17-23, 1970. Edited by H. Dinges. V, 75 pages. 1971. DM 16,–

Vol. 191: Séminaire de Probabilités V. Edited by P. A. Meyer. IV, 372 pages. 1971. DM 26,–

Vol. 192: Proceedings of Liverpool Singularities – Symposium I. Edited by C. T. C. Wall. V, 319 pages. 1971. DM 24,–

Vol. 193: Symposium on the Theory of Numerical Analysis. Edited by J. Ll. Morris. VI, 152 pages. 1971. DM 16,–

Vol. 194: M. Berger, P. Gauduchon et E. Mazet. Le Spectre d'une Variété Riemannienne. VII, 251 pages. 1971. DM 22,–

Vol. 195: Reports of the Midwest Category Seminar V. Edited by J.W. Gray and S. Mac Lane. III, 255 pages. 1971. DM 22,–

Vol. 196: H-spaces – Neuchâtel (Suisse)- Août 1970. Edited by F. Sigrist, V, 156 pages. 1971. DM 16,–

Vol. 197: Manifolds – Amsterdam 1970. Edited by N. H. Kuiper. V, 231 pages. 1971. DM 20,–

Vol. 198: M. Hervé, Analytic and Plurisubharmonic Functions in Finite and Infinite Dimensional Spaces. VI, 90 pages. 1971. DM 16.–

Vol. 199: Ch. J. Mozzochi, On the Pointwise Convergence of Fourier Series. VII, 87 pages. 1971. DM 16,–

Vol. 200: U. Neri, Singular Integrals. VII, 272 pages. 1971. DM 22,–

Vol. 201: J. H. van Lint, Coding Theory. VII, 136 pages. 1971. DM 16,–

Vol. 202: J. Benedetto, Harmonic Analysis on Totally Disconnected Sets. VIII, 261 pages. 1971. DM 22,–

Vol. 203: D. Knutson, Algebraic Spaces. VI, 261 pages. 1971. DM 22,–

Vol. 204: A. Zygmund, Intégrales Singulières. IV, 53 pages. 1971. DM 16,–

Vol. 205: Séminaire Pierre Lelong (Analyse) Année 1970. VI, 243 pages. 1971. DM 20,–

Vol. 206: Symposium on Differential Equations and Dynamical Systems. Edited by D. Chillingworth. XI, 173 pages. 1971. DM 16,–

Vol. 207: L. Bernstein, The Jacobi-Perron Algorithm – Its Theory and Application. IV, 161 pages. 1971. DM 16,–

Vol. 208: A. Grothendieck and J. P. Murre, The Tame Fundamental Group of a Formal Neighbourhood of a Divisor with Normal Crossings on a Scheme. VIII, 133 pages. 1971. DM 16,–

Vol. 209: Proceedings of Liverpool Singularities Symposium II. Edited by C. T. C. Wall. V, 280 pages. 1971. DM 22,–

Vol. 210: M. Eichler, Projective Varieties and Modular Forms. III, 118 pages. 1971. DM 16,–

Vol. 211: Théorie des Matroïdes. Edité par C. P. Bruter. III, 108 pages. 1971. DM 16,–

Please turn over

Vol. 212: B. Scarpellini, Proof Theory and Intuitionistic Systems. VII, 291 pages. 1971. DM 24,-

Vol. 213: H. Hogbe-Nlend, Théorie des Bornologies et Applications. V, 168 pages. 1971. DM 18,-

Vol. 214: M. Smorodinsky, Ergodic Theory, Entropy. V, 64 pages. 1971. DM 16,-

Vol. 215: P. Antonelli, D. Burghelea and P. J. Kahn, The Concordance-Homotopy Groups of Geometric Automorphism Groups. X, 140 pages. 1971. DM 16,-

Vol. 216: H. Maaß, Siegel's Modular Forms and Dirichlet Series. VII, 328 pages. 1971. DM 20,-

Vol. 217: T. J. Jech, Lectures in Set Theory with Particular Emphasis on the Method of Forcing. V, 137 pages. 1971. DM 16,-

Vol. 218: C. P. Schnorr, Zufälligkeit und Wahrscheinlichkeit. IV, 212 Seiten 1971. DM 20,-

Vol. 219: N. L. Alling and N. Greenleaf, Foundations of the Theory of Klein Surfaces. IX, 117 pages. 1971. DM 16,-

Vol. 220: W. A. Coppel, Disconjugacy. V, 148 pages. 1971. DM 16,-

Vol. 221: P. Gabriel und F. Ulmer, Lokal präsentierbare Kategorien. V, 200 Seiten. 1971. DM 18,-

Vol. 222: C. Meghea, Compactification des Espaces Harmoniques. III, 108 pages. 1971. DM 16,-

Vol. 223: U. Felgner, Models of ZF-Set Theory. VI, 173 pages. 1971. DM 16,-

Vol. 224: Revètements Etales et Groupe Fondamental. (SGA 1). Dirigé par A. Grothendieck XXII, 447 pages. 1971. DM 30,-

Vol. 225: Théorie des Intersections et Théorème de Riemann-Roch. (SGA 6). Dirigé par P. Berthelot, A. Grothendieck et L. Illusie. XII, 700 pages. 1971. DM 40,-

Vol. 226: Seminar on Potential Theory, II. Edited by H. Bauer. IV, 170 pages. 1971. DM 18,-

Vol. 227: H. L. Montgomery, Topics in Multiplicative Number Theory. IX, 178 pages. 1971. DM 18,-

Vol. 228: Conference on Applications of Numerical Analysis. Edited by J. Ll. Morris. X, 358 pages. 1971. DM 26,-

Vol. 229: J. Väisälä, Lectures on n-Dimensional Quasiconformal Mappings. XIV, 144 pages. 1971. DM 16,-

Vol. 230: L. Waelbroeck, Topological Vector Spaces and Algebras. VII, 158 pages. 1971. DM 16,-

Vol. 231: H. Reiter, L^1-Algebras and Segal Algebras. XI, 113 pages. 1971. DM 16,-

Vol. 232: T. H. Ganelius, Tauberian Remainder Theorems. VI, 75 pages. 1971. DM 16,-

Vol. 233: C. P. Tsokos and W. J. Padgett. Random Integral Equations with Applications to Stochastic Systems. VII, 174 pages. 1971. DM 18,-

Vol. 234: A. Andreotti and W. Stoll. Analytic and Algebraic Dependence of Meromorphic Functions. III, 390 pages. 1971. DM 26,-

Vol. 235: Global Differentiable Dynamics. Edited by O. Hájek, A. J. Lohwater, and R. McCann. X, 140 pages. 1971. DM 16,-

Vol. 236: M. Barr, P. A. Grillet, and D. H. van Osdol. Exact Categories and Categories of Sheaves. VII, 239 pages. 1971, DM 20,-

Vol. 237: B. Stenström. Rings and Modules of Quotients. VII, 136 pages. 1971. DM 16,-

Vol. 238: Der kanonische Modul eines Cohen-Macaulay-Rings. Herausgegeben von Jürgen Herzog und Ernst Kunz. VI, 103 Seiten. 1971. DM 16,-

Vol. 239: L. Illusie, Complexe Cotangent et Déformations I. XV, 355 pages. 1971. DM 26,-

Vol. 240: A. Kerber, Representations of Permutation Groups I. VII, 192 pages. 1971. DM 18,-

Vol. 241: S. Kaneyuki, Homogeneous Bounded Domains and Siegel Domains. V, 89 pages. 1971. DM 16,-

Vol. 242: R. R. Coifman et G. Weiss, Analyse Harmonique Non-Commutative sur Certains Espaces. V, 160 pages. 1971. DM 16,-

Vol. 243: Japan-United States Seminar on Ordinary Differential and Functional Equations. Edited by M. Urabe. VIII, 332 pages. 1971. DM 26,-

Vol. 244: Séminaire Bourbaki - vol. 1970/71. Exposés 382-399. IV, 356 pages. 1971. DM 26,-

Vol. 245: D. E. Cohen, Groups of Cohomological Dimension One. V, 99 pages. 1972. DM 16,-

Vol. 246: Lectures on Rings and Modules. Tulane University Ring and Operator Theory Year, 1970-1971. Volume I. X, 661 pages. 1972. DM 40,-

Vol. 247: Lectures on Operator Algebras. Tulane University Ring and Operator Theory Year, 1970-1971. Volume II. XI, 786 pages. 1972. DM 40,-

Vol. 248: Lectures on the Applications of Sheaves to Ring Theory. Tulane University Ring and Operator Theory Year, 1970-1971. Volume III. VIII, 315 pages. 1971. DM 26,-

Vol. 249: Symposium on Algebraic Topology. Edited by P. J. Hilton. VII, 111 pages. 1971. DM 16,-

Vol. 250: B. Jónsson, Topics in Universal Algebra. VI, 220 pages. 1972. DM 20,-

Vol. 251: The Theory of Arithmetic Functions. Edited by A. A. Gioia and D. L. Goldsmith VI, 287 pages. 1972. DM 24,-

Vol. 252: D. A. Stone, Stratified Polyhedra. IX, 193 pages. 1972. DM 18,-

Vol. 253: V. Komkov, Optimal Control Theory for the Damping of Vibrations of Simple Elastic Systems. V, 240 pages. 1972. DM 20,-

Vol. 254: C. U. Jensen, Les Foncteurs Dérivés de $\varprojlim$ et leurs Applications en Théorie des Modules. V, 103 pages. 1972. DM 16,-

Vol. 255: Conference in Mathematical Logic - London '70. Edited by W. Hodges. VIII, 351 pages. 1972. DM 26,-

Vol. 256: C. A. Berenstein and M. A. Dostal, Analytically Uniform Spaces and their Applications to Convolution Equations. VII, 130 pages. 1972. DM 16,-

Vol. 257: R. B. Holmes, A Course on Optimization and Best Approximation. VIII, 233 pages. 1972. DM 20,-

Vol. 258: Séminaire de Probabilités VI. Edited by P. A. Meyer. VI, 253 pages. 1972. DM 22,-

Vol. 259: N. Moulis, Structures de Fredholm sur les Variétés Hilbertiennes. V, 123 pages. 1972. DM 16,-

Vol. 260: R. Godement and H. Jacquet, Zeta Functions of Simple Algebras. IX, 188 pages. 1972. DM 18,-

Vol. 261: A. Guichardet, Symmetric Hilbert Spaces and Related Topics. V, 197 pages. 1972. DM 18,-

Vol. 262: H. G. Zimmer, Computational Problems, Methods, and Results in Algebraic Number Theory. V, 103 pages. 1972. DM 16,-

Vol. 263: T. Parthasarathy, Selection Theorems and their Applications. VII, 101 pages. 1972. DM 16,-

Vol. 264: W. Messing, The Crystals Associated to Barsotti-Tate Groups: with Applications to Abelian Schemes. III, 190 pages. 1972. DM 18,-

Vol. 265: N. Saavedra Rivano, Catégories Tannakiennes. II, 418 pages. 1972. DM 26,-

Vol. 266: Conference on Harmonic Analysis. Edited by D. Gulick and R. L. Lipsman. VI, 323 pages. 1972. DM 24,-

Vol. 267: Numerische Lösung nichtlinearer partieller Differential- und Integro-Differentialgleichungen. Herausgegeben von R. Ansorge und W. Törnig, VI, 339 Seiten. 1972. DM 26,-

Vol. 268: C. G. Simader, On Dirichlet's Boundary Value Problem. IV, 238 pages. 1972. DM 20,-

Vol. 269: Théorie des Topos et Cohomologie Etale des Schémas. (SGA 4). Dirigé par M. Artin, A. Grothendieck et J. L. Verdier. XIX, 525 pages. 1972. DM 50,-

Vol. 270: Théorie des Topos et Cohomologie Etle des Schémas. Tome 2. (SGA 4). Dirige par M. Artin, A. Grothendieck et J. L. Verdier. V, 418 pages. 1972. DM 50,-

Vol. 271: J. P. May, The Geometry of Iterated Loop Spaces. IX, 175 pages. 1972. DM 18,-

Vol. 272: K. R. Parthasarathy and K. Schmidt, Positive Definite Kernels, Continuous Tensor Products, and Central Limit Theorems of Probability Theory. VI, 107 pages. 1972. DM 16,-

Vol. 273: U. Seip, Kompakt erzeugte Vektorräume und Analysis. IX, 119 Seiten. 1972. DM 16,-

Vol. 274: Toposes, Algebraic Geometry and Logic. Edited by. F. W. Lawvere. VI, 189 pages. 1972. DM 18,-

Vol. 275: Séminaire Pierre Lelong (Analyse) Année 1970-1971. VI, 181 pages. 1972. DM 18,-

Vol. 276: A. Borel, Représentations de Groupes Localement Compacts. V, 98 pages. 1972. DM 16,-

Vol. 277: Séminaire Banach. Edité par C. Houzel. VII, 229 pages. 1972. DM 20,-